To My Parents

JOSEPH AND ELIYAMMA KOZHAMTHADAM

CONTENTS

ACKNOWLEDGMENTS

I want to express my sincere gratitude to a number of persons who have made this work possible. First of all, my deepest gratitude goes to Dudley Shapere. His inspiring lectures at the University of Maryland were instrumental in instilling in me a serious interest in Kepler. Over the years he has been a genuine source of help, inspiration, and encouragement. My special thanks are due also to Frederick Suppe for his personal interest, encouragement, and valuable suggestions at various stages of this work. Stephen Brush read its draft in the initial stages and made many valuable suggestions; I am grateful to him for this and for all the help and encouragement he has given so generously.

I want to thank the late Francis Haber, Owen Gingerich, Eugene Helm, James Langford, and Curtis Wilson for reading the manuscript and giving their expert comments and advice.

It is well known that Kepler's original writings are very difficult to translate. I have found helpful the existing translations of Carola Baumgardt, William Donahue, A. M. Duncan, Owen Gingerich, Alexandre Koyré, Wolfgang Pauli, Charles Wallis, and Ann Wegner. I have modified or retranslated the relevant passages whenever it was necessary to bring out more clearly the meaning of the original Latin text. In this connection I am particularly grateful to Owen Gingerich for his generously making available to me his collection of draft translations of Kepler's writings.

Robert Harvanek, S.J., and the Jesuit Community of the Loyola University of Chicago generously provided funds for the last phase of this work. Similarly Joseph Übelmesser, S.J., the Mission Procurator of the German Jesuit Province, also helped this project with generous financial assistance. I express my deep appreciation and gratitude to them.

My special thanks also go to many friends and colleagues, particularly to Dr. Stella Mathew, for their inspiration and support all through the course of this research.

Finally, I am grateful to Jeannette Morgenroth for her excellent editorial help and to the University of Notre Dame Press for publishing this book.

ABBREVIATIONS

AN = *Astronomia Nova*
BB = Arthur Beer and Peter Beer, *Kepler: Four Hundred Years*
CB = Carola Baumgardt, *Johannes Kepler: Life and Letters*
DR = *De Revolutionibus*
ER = Edward Rosen's translation of *De Revolutionibus*
GR = *Great Books of the Western World*
HM = *Harmonices Mundi*
KY = Alexandre Koyré, *The Astronomical Revolution*
MC = *Mysterium Cosmographicum*
RS = Robert Small, *An Account of the Astronomical Discoveries of Kepler*
SM = *Sidereal Messenger* 6 (1887)

INTRODUCTION

Despite having extreme interest and importance, the nature of scientific discovery has always been elusive and has remained an unsolved philosophico-scientific problem. How do scientists make their discoveries? Is there a logic of scientific discovery? If so, what is it? If not, then how can scientific discovery differ from luck or witchcraft? Again, what factors are involved in scientific discovery? Do extrascientific concerns, especially philosophical and religious principles, have a role to play? What specifically is that role? A detailed and in-depth study of Kepler's discovery of the first two laws of planetary motion, this book attempts to throw light on the above questions and related ones.

Perhaps no other case of scientific discovery is so interesting and enlightening as that of Kepler's laws. This is so not only because his laws are among the first ones discovered in modern science, but also because he has left extensive and candid[1] reports on the various paths he took in his journey toward these laws (except for the third law).

A scientific discovery can be divided into two stages or steps: (1) the initial "thinking up" of the idea, and (2) the eventual establishment of the idea. The present study examines both stages. According to some writers, the first step cannot have a logic or rationality, which seems to imply that hypotheses originate in a rational vacuum. Many contemporary philosophers of science oppose such a view, and we shall see that Kepler himself already in the 1600s had argued against it. Other scholars consider the first stage to be the real discovery because there the flash of genius is often most conspicuous. However, I believe that such an understanding of discovery is insufficient because very often the initial insights and ideas turn out to be scientifically trivial or even false. Only when the initial idea has been established satisfactorily can we say that a genuine and fruitful scientific theory has been discovered, and that process is not to be seen as completely separate from discovery but, at least for Kepler, as its culmination.

Attempts to unravel Kepler's discovery have led to a wide spectrum of views. Some argue that he did not discover any truly physical laws but only mathematical regularities.[2] It was rather the immortal Newton who actually discovered the laws and who gave the physical explanation for them. Newton himself argued that Kepler got the laws by lucky guesswork and therefore did not deserve credit for their discovery (that honor Newton reserved for himself). Arthur Koestler[3] would give full credit to Kepler for the discovery but with an important qualification: the discovery was achieved not by a careful, systematic, logically reasoned, and empirically tested process but, rather, by a strange, unpredictable process of "sleepwalking," with which rationality had very little to do and in which the discoverer basically was groping in the dark, not knowing exactly what he was doing. On the other hand, a common view, not adopted by recent scholars, is that Kepler arrived at the laws by "curve-fitting." This view holds that he, using the accurate data provided by Tycho (the observational genius of the century), plotted the different points that the planet Mars occupied at various positions of its path and thus found the exact path it followed. Scholars holding this position seem to believe that the laws were almost exclusively the result of the empirical investigations of Kepler and Tycho.[4] Some more recent scholars who take a middle position argue that Kepler's discovery resulted from following up certain "hunches" by means of empirical tests.[5] It seems to me that all these views are either incorrect or incomplete because they fail to take into account all of the important factors that actively shaped Kepler's work.

Different views also exist concerning the logic of Kepler's discovery. I have already referred to Koestler's position. At the other extreme, Norwood Russell Hanson, following the lead of Charles

Peirce, believes that there was a definite logic, which Hanson identifies as retroduction or abduction. In fact, quoting Peirce, he says that Kepler's discovery "is the greatest piece of retroductive reasoning ever performed."[6] I shall point out in my concluding remarks that Hanson's retroductive analysis of Kepler's discovery is too simplistic to be of much help. The process of Keplerian discovery was certainly rational, but so far no attempt to identify a methodical and well-defined logic for it has been successful.

With regard to the role that nonempirical or nonscientific factors, especially philosophical and religious principles, played in this discovery, a variety of views have emerged. Small and Dreyer do not even deem such a question worth considering since they seem to take for granted that there were practically no such influences on Kepler. Some others consider these factors, only to deny them any positive role;[7] these scholars seem to presuppose that Kepler had adequate empirical evidence for making his scientific discoveries. This position, I believe, is unwarranted and incorrect. In many significant respects, empirical evidence was not adequate to arrive at the conclusions that Kepler actually did draw. Further considerations of nonempirical, and, indeed, of philosophical and religious nature entered into the process of his scientific discovery in crucial ways. Some other scholars are willing to allot a significantly limited role to these nonscientific factors.[8] Although providing valuable insights into Kepler's original work, their discussion of these factors remains very underdeveloped and incomplete. For instance, Wilson emphasizes the role of "hunches": Kepler's discovery required more than scientific factors, at least if we consider these "hunches" to be nonscientific. But a serious and adequate discussion of their source, nature, and role in the process of discovery is left out. Why did Kepler come up with these specific hunches rather than some other ones? What, if any, "logic" or "rationale" led to them? These and related questions remain unanswered.

Alexandre Koyré clearly assigns a definite role to nonempirical factors in the discovery of the laws. In his view, Kepler's belief in the mathematical nature of God, humankind, and the universe was crucial for the astronomer's success. Above all, the secret of his success consisted in this: "Kepler asked himself—*a quo moventur planetae* (what is it that makes the planets move)? Kepler gave this question a dynamic significance, which was something no one had done before."[9] However, Koyré neither identifies Kepler's different religious beliefs, apart from a rather vague reference to his belief in the mathematical nature of God, nor analyses exactly how Kepler's beliefs influenced his thinking and scientific work. Although Koyré talks about some of Kepler's philosophical beliefs, the scholar leaves out several important ones. Furthermore, he does not give any detailed or systematic study of these beliefs or their precise roles in the discovery of the laws. Hence Koyré's study fails to do full justice to the influence of philosophy and, especially, religion on Kepler's discovery.

In contrast to the scholars discussed above, some others overemphasize philosophical and religious influences on Kepler and his work. Edwin Burtt and Arthur Koestler belong to this group.[10] According to Burtt, Kepler was basically a "sun worshipper" and this was the "main sufficient reason" for his total acceptance of heliocentrism. This view, as we shall see in part 2, does not square with the facts. Koestler also brings out the importance of the religious elements in Kepler's work but, as I have mentioned earlier, at the expense of seeing no rationality in the process of discovery. My own investigations lead me to conclude that each and every step that brought Kepler to success was the result of careful reasoning on the basis of the various ideas and data available to him.

Gerald Holton, in his interesting and insightful paper, argues that nonscientific factors, specifically philosophical and religious principles, played a role in Kepler's discovery. More than anybody else Holton brings out the threefold source of Kepler's thought and makes some attempt to identify their elements. "In one brilliant image, Kepler saw the three basic cosmological models superposed: the universe as physical machine, the universe as mathematical harmony, and the universe as central theological order."[11] Holton correctly acknowledges that the three different factors actively guided Kepler to discovery. Also our attention is accurately and correctly called to the "apparent confusion of incongruous elements—physics and metaphysics, astronomy and astrology, geometry and theology—which characterizes Kepler's work."[12] However, it seems to me that, even though Holton's overall picture of Kepler's thought is more encompassing than others', it nevertheless remains sketchy, rarely going beyond being broadly thematic and suggestive. For instance, towards the end of the paper he says: "A theocentric conception of the universe led to specific results of crucial importance for the rise of modern physics."[13] But he leaves undeveloped how these specific results were reached. Even though Holton's general observations about Kepler have much to offer, we need a more solid and detailed exposition of the ideas involved in each of the three factors—religion, philosophy, and science—and how they shaped Kepler's thought. The present work attempts to accomplish this.

One needs to reconsider also the nature of the interaction that Holton envisages. According to him, the interaction was essentially a rescue operation. As he puts it: When Kepler's "physics fails, his metaphysics comes to the rescue; when the mechanical model is insufficient as a tool of explanation, a mathematical model takes over, and at its boundary in turn there stands a theological axiom."[14] This description implies that the process of discovery was a rescue operation for Kepler's system when his physics was in deep trouble. For Holton, philosophy and religion take over only when physics fails. Thus these two other areas form a rescue squad rather than being respectable partners in a common mission. Further, according to Holton, religion was the least significant; it had only a marginal importance since "it stood *at the boundary.*" In my opinion, this does not seem to be what we see accurately in Kepler's writings. Some part of the AN may convey this impression, but then this was not Kepler's only work; nor did he deem it the most important one. To complete a view of Kepler we need to consider the other works as well, especially the MC and HM: Kepler considered the former to be the origin and source of all his ideas, and the latter, the culmination of his intellectual striving. I think that his other scientific works, though they add to the total picture, must be seen in the light of his overall thought as expressed largely in these books. When we have a more complete image of Kepler, the picture of respectable partnership emerges.

Nor should philosophical and religious principles be considered merely a temporary scaffolding; not meant to fill in for a time and then disappear from his work, they were present and active all through his long journey to the laws. Again, philosophical and the religious ideas did not play a role only as a framework, for they were more central and substantive than that; the ideas were actually and directly responsible for discovery, unlike a framework, where an idea only contributes to the creation of the appropriate attitude and atmosphere conducive to scientific discovery. For instance, we know that dedication and perseverance are often needed for scientific discovery. If religious principles motivate a person to have these qualities, we say that religion is playing a framework role. Although at times philosophy and religion played a framework role for Kepler, their contribution was on the whole substantive.

My principal thesis is that Kepler's discovery was the result of the interaction among his empirical (scientific), philosophical, and religious ideas. All three had to collaborate to make the discovery possible. Each had a specific contribution to make that the other two could not have made. To be sure, the uniqueness of the contribution of each area was more striking in some stages of his discovery than in some others, but it was there always. If he had emphasized only the empirical, he probably would have been another Tycho, but not the discoverer of the laws and the father of modern astronomy. On the other hand, if he had concentrated only on the philosophical, he would have remained just one among many Renaissance natural philosophers. Religious ideas alone also would not have led him to the laws. But when all three could interact and collaborate with each other, mutually modifying and complementing each other, the result was the laws.

This study has two parts. The first part will outline the general tenets of Kepler's religious view and his main philosophical principles; it will examine also his scientific ideas and their relations to his religious and philosophical thought. The second part will discuss how all the three areas worked together to lead him to the second and the first laws. I divide the discovery of the two laws into different steps or stages and show how in each stage all three factors made their specific contributions.

Coming to the methodology adopted in this book, I would emphasize that it is a historico-philosophical study. Hence serious attempt has been made to develop the various ideas in the light of Kepler's own original works and the critical analysis of them. Although the first three chapters discuss his religious, philosophical, and scientific views, it is not an exhaustive study of these three aspects.[15] In each of these three chapters we shall focus only on those ideas and principles that had a direct bearing on the process of the discovery of the laws.

A few words about the kind of evidence presented for the interaction of religion, philosophy, and science may not be out of place here. Although direct textual evidence will be given for the tripartite partnership in all the important stages of the discovery, in some stages it is more striking than in some others. One may remark that towards the final stages of the process of discovery, the interaction was not so conspicuous, especially with regard to the role of religious principles. As I shall point out in detail in part 2, this does not mean that the religious dimension was absent but, rather, that it was taken for granted. It seems that Kepler was so convinced of the obviousness of the role of God and religion that he did not feel it necessary to explicitly express it in AN. In his letters we do find that religious considerations were very active in his work during the highpoint of the final stages of his discovery. For instance, in his letter to Maestlin, written only a few months before he announced to Fabricius the discovery of the ellipse, Kepler explicitly acknowledged the assistance of God in the discovery of the noncircular orbit.[16] Again, in a letter to Herwart, written even later and so still closer to the all-important announcement of the ellipse, Kepler once again emphasized his favorite themes of the Trinity—sphere analogy and the supreme importance of the sun.[17] Similar statements can be found just after his announcement and in his later years.[18] We have reliable textual evidence for the influence of religion on his thought and work before, during, and after his work on Mars.

What I mean by *philosophical* will become clear only later. I will explain it not so much by defining the term as by describing what its principal elements are and how they function in Kepler's work.

I will have some things to say about the general question of whether the role of nonscientific factors, so prominent in Kepler, is a general feature of all science. However, it will not be my purpose to study in detail the question of whether philosophical and religious ideas are always necessary for scientific discovery. My central concern is with Kepler. Although I will point out at length that a genuine rational process was involved at every step of Kepler's discovery, I shall not attempt to develop a systematic logic of discovery and thus will not offer complete answers to the questions that I posed in the first paragraph of this introduction. But I hope that this case study will throw considerable light on them, as I will point out in the conclusion. There, in the light of the new interpretation, I shall also reassess Kepler's importance in the histories of religion or theology, philosophy, and science. This investigation will also reveal that Kepler was truly a transitional figure who contributed significantly in the passage from medieval to modern science.

I have left out any discussion of the discovery of the third law for two reasons: (1) There is very little information about it in Kepler's own writings. Although he knew the problem (as did Copernicus himself) even in MC, in HM he suddenly announced this law without providing any clear clue as to how he had arrived at it. (2) The general points I wish to make will be brought out adequately in relation to the first two laws and so a discussion of the third is not necessary for my purpose. I hope this study will contribute not only to Kepler scholarship but also to the more general philosophical issue of the "logic" of scientific discovery.

Part 1

Kepler's System of Thought

1

KEPLER'S RELIGIOUS IDEAS

The rational God, who does nothing in vain, whose actions, though at times inscrutable for us humans, are never without a reason, who self-manifests in the universe through discernible, orderly laws of nature based on geometry and harmony, forms one segment of the circle of Kepler's religious view of the universe. The human being, the image of God, commissioned to give praise and honor to the supreme Deity by uncovering and making known those laws, constitutes the other segment. The circle consists in God's coming down to engage humans and their going up to embrace God, a movement mediated through nature. As Kepler put it: "What voice has the heaven, what voice have the stars, to praise God as man does? Unless, when they supply man with cause to praise God, they themselves are said to praise God."[1] Since this encounter takes place in and through nature, the universe is an integral part of this meeting. The universe and the study of it cannot be divorced from Kepler's religion. Thus he could write to David Fabricius in 1603: "For me nature aspires to divinity."[2] God is supremely rational and the human being, God's image and likeness, cannot but share in this rationality. Indeed, one of the main reasons why the human being is the image of God is precisely because of being rational. Religion, therefore, for Kepler had to be permeated by rationality. Hence God, the human being, the universe, and rationality formed the tetrad that built up the edifice of Kepler's life as a religious believer. All the various elements of religion in this deeply religious person could in turn be woven into this tetrad.

Kepler the religious believer was a transitional figure through and through. Elements of traditional and medieval religious views are to be seen in his ideas and practices. On the other hand, he also held views radically different from the traditional. This chapter and the next two will discuss the traditional outlook as well as his own original views so as to bring out the creativity and transitional character of his contributions.

In Kepler's thought process, the religious, the philosophical, and the empirical were intimately related. One may say that his basic unit of thought was "interconnected" categories. His thought pattern was of the "interlocking" type. In order to identify these thought patterns, in our study of Kepler we shall examine the mutual interaction among these three aspects and the influence of each on the other.

THE KEPLERIAN IDEA OF GOD[3]

The Unique Place of God in Kepler's System

A religious belief or view, centered around and dependent on a supreme being, necessarily has a direct reference to the one supreme God or, in polytheism, to many gods.[4] The unique role attributed to God distinguishes religious belief from all other forms of belief. Imbued with a God-originated, God-guided, and God-oriented view, Kepler believed that God was the origin, source, direction, and sole goal of all his thought and works in every detail. His was a totally theocentric system. Max Caspar, the greatest Kepler scholar of our century, expresses this idea: "God is truth, and service to truth proceeds from him and leads to him. God is the beginning and end of his scientific research and striving. Therein lies the keynote of Kepler's thought, the basic motive of his purpose, and the life-giving soil of his feeling."[5] One can hardly scan a page of his works (except when he is discussing purely technical points and calculations) without encountering his mentioning God, at least indirectly.

Several of his major works (especially works announcing his most important discoveries) begin and end with a prayer of thanksgiving and praise to the Almighty. For instance, when he broke the news to Michael Maestlin about the very first discovery of the polyhedral theory,[6] Kepler wrote: "God who is the most admirable in his works may deign to grant us the grace to bring to light and illuminate the profundity of his wisdom in the visible (and accordingly intelligible) creation of this world."[7]

Thus exceedingly excited though he was about this discovery, he took it not so much as a personal victory or achievement as an occasion to admire "the profundity" of God's "wisdom." In the beginning of MC, in the "Preface to the Reader," the very first sentence announced that his intention was only to show how the greatest and wisest creator carried out the creation of our moving universe.[8] He wanted to make clear at the very outset that his effort simply unraveled and made public the plan of God. Hence it is no wonder that this new David ended his first major book with a beautiful hymn to the Lord and Master of Creation, which is nothing but a paraphrase of Psalm 8 (where the Psalmist extols and enumerates the wonders wrought by the wisest one). He prefaced the final hymn as follows:

> Now, friendly reader, do not forget the end of all this, which is the conception, admiration and veneration of the Most Wise Maker. For it is nothing to have progressed from the eyes to the mind, from sight to contemplation, from the visible motion to the Creator's most profound plan, if you are willing to rest there, and do not soar in a single bound and with complete dedication of spirit to knowledge, love and worship of the Creator. . . .[9]

This passage and the hymn following it showed the central role God played in his work. Again, in his introduction to AN, his most scientific and highly technical book, he invited the reader to join him

> to praise and glorify with me the wisdom and greatness of the Creator, which I have revealed in a deeper explication of the form of the universe, in an investigation of the causes, and in my detection of the deceptiveness of sight. Let him not only extol the bounty of God in the preservation of living creatures of all kinds by the strength and stability of the earth, but also let him acknowledge the wisdom of the creator in its motion, so abstruse, so admirable.[10]

According to Kepler, the whole laborious work he accomplished in the AN was directed to the praise and glory of God; this sublime goal made the work valuable and worthwhile. The HM he considered the highest point of his life, the fulfillment of his life ambition. He made no secret of this fact when he wrote concerning what he had accomplished through this book: "The purpose which drove me to spend the best part of my life in studying astronomy, to seek out Tycho Brahe, and to choose Prague as my home, that with God's help I have finally fulfilled."[11] The same sentiment pervaded this book. In the very beginning of the most important part of this work (book 5) he quoted Galen, obviously making the words his own: "I commence a sacred discourse, a most true hymn to God the Founder, and I judge it to be piety, not to sacrifice many hecatombs of bulls to Him and to burn incense of innumerable perfumes and Cassia, but first to learn myself, and afterwards to teach others too, how great He is in wisdom, how great in power, and of what sort in goodness."[12] In the same introduction he affirmed that the "finger of God" was guiding him in this discovery, and in an exuberant frenzy he cried out: "I am free to give myself to the sacred madness, I am free to taunt mortals with the frank confession that I am stealing the golden vessels of the Egyptians, in order to build of them a temple for my God, far from the territory of Egypt."[13] Once more the purpose of all these years of toil was to build a fit temple for his God. Naturally he concluded this long book by once again imitating the royal Psalmist: "Great is our Lord and great his power, and of [the works of] His wisdom there is no number . . . for from him, through him, and in him are all things both the sensible and the intelligible. . . . To him be praise, honor and glory, world without end, Amen."[14] Much can be said about this conclusion, nearly all of it beyond the scope of my present work; so I shall only refer to a few more detailed studies on this hymn.[15] Caspar points out that the very last sentence in Kepler's last letter was: "Hold fast with confidence along with me to the only anchor of the Church, prayers to God for it and for me."[16]

The passages above paint a portrait of a person who was intoxicated with the idea and experience of God. But I will go further to suggest that the idea and the felt experience of God was crucial and substantive in all his works and thoughts, in his whole system. Indeed, God provided him with one of the principal ways of explaining physical reality. For instance, he wrote in MC: "I think that from the love of God for man a great many of the causes and of the features in the universe can be deduced."[17] God was no intruder from outside, nor did God enter Kepler's works as an afterthought cunningly introduced to placate some fanatic censor. The often-quoted and often-imitated Laplacian claim that a scientific system has no need of God would have been utter nonsense for Kepler. Holton suggests this same idea: "In the end, Kepler's unifying principle for the world of phenomena is not merely the conception of mathematical forces, but God. . . ."[18] Indeed the idea of God was fundamental for Kepler because on it was based his understanding of the universe and the laws governing it. Science also became purposeful and meaningful because of its relation to God. In short, Kepler's concept of God was at the root of his science, cosmology, metaphysics, epistemology, and philosophy of man. Furthermore, as we shall see in the next section, his idea of God differed significantly from that of his contemporaries and, in some ways, was unique to him.

Nature of God

To establish my claim that Kepler's conception of God had a crucial substantive place in his system, a detailed discussion of his understanding of God is necessary. Although he shared traditional ideas, he interpreted them in a distinctly different way. This interpretation was the result of the co-integrative influence of his philosophical and empirical views. In turn, this new way of understanding molded his ideas in the other two areas so as to make a positive and crucial contribution to his philosophy and to the discovery of scientific laws.

GOD THE "OPTIMUS CREATOR"

The belief that God is the creator of the universe is as old as religion itself. All major religions consider God as the creator of the world. Kepler subscribed to this view wholeheartedly. Further, the best and wisest God has created the best world. "For it neither is nor was right that he who is the best should make anything except the most beautiful."[19] According to him, in this world of ours "nothing is in excess, nothing is in want, there is no room for any censure."[20] In this regard he seemed to have anticipated the Leibnizean optimism, but for different reasons. In Kepler's view our universe is the best because on it is imprinted the most perfect thing of all: God's own essence. He wrote in MC:

> Since then the creator conceived the Idea of the universe in his mind (we speak in human fashion, so that being men we may understand), and it is the Idea of that which is prior, indeed, as has just been said, of that which is best, so that the Form of the future creation may itself be the best: it is evident that by those laws which God himself in his goodness prescribes for himself, the only thing of which he could adopt the Idea for establishing the universe is his own essence. . . .[21]

Although he warned his reader that he was talking in human terms, his position was far from being a figure of speech. He did mean that the universe had the imprint of the essence of God, that the universe had been created in accordance with the very essence of God. Later he would argue that this relationship between the essence of God and the universe had been brought about through geometry. Geometry is part of the essence of God, and the world shares in God's essence insofar as it has been created according to the canons of geometry. Thus God, geometry, and nature were intimately interconnected in Kepler's thought. From this it follows that his idea of geometry and of the structure and nature of the world were based on his idea of God.

GOD THE TRINITY

Belief in the Holy Trinity has been an integral part of the Christian tradition, despite its unintelligibility and incomprehensibility. Although Kepler had serious reservations about several other Christian doctrines (e.g., the real presence of Christ in the eucharist, Calvinist predestination, infallibility, etc.), he was an ardent believer in the mystery of the Holy Trinity. He believed that only one God exists, but three persons are in the Godhead: the Father, the Son, and the Holy Spirit. The Son proceeds from the Father, and the Spirit from the Father and the Son. Each Person is distinct, although the three together constitute the one true God. While he refused to give any special significance to pure numbers as such, the number three had a special and privileged position because of the trinitarian character of the Deity.

Kepler related the Trinity to the universe and in the process attempted to make the Trinity less mysterious. The trinitarian God he compared to a sphere: the Father representing the center; the Son, the surface; and the Spirit, the intervening space. Nicholas of Cusa before him had made a similar comparison, but Kepler emphasized this comparison far more and, indeed, made it an integral part of his system. According to Hübner, the Trinity–sphere idea was at the very foundation of Kepler's thinking: "Obviously here we have to do with a basic idea of his thought, which decidedly defines his worldview."[22] As we shall see later, Kepler would emphasize that this idea was no mere metaphor, that it really represented the actual structure of the universe.

Reference to this comparison occurred in so many of his writings that one senses in him an obsession with it. In 1595 he mentioned it in his letter to Maestlin,[23] in 1596 we see it in the MC,[24] in 1604 in the *Astronomia Pars Optica*,[25] in 1605 in the letter to Herwart,[26] in 1608 in the letter to Tankius,[27] in 1610 in *Tertius Interveniens*,[28] in 1618 in the *Epitome*,[29] in 1620 in HM.[30] This metaphor turns up in many places, in many different kinds of books, spanning many years. One wonders why he was so much obsessed with the Trinity-as-sphere. I would suggest a number of reasons for this extraordinary interest. Kepler accepted the Holy Trinity as a mystery, as something that could never fully be fathomed, could only be approached but never be fully solved. He wanted to get as close to it as possible by making it as intelligible as possible. The emphasis on this analogy highlighted the importance he attached to intelligibility. Even an unfathomable mystery he wanted to render intelligible.[31] Secondly, the geometrical nature of the analogy must have exerted a strong attraction. The sphere is the most perfect of all bodies and hence what better candidate could be found for a comparison with the most sublime Trinity? This analogy also reaffirmed his optimism about the nature of the world. If the universe can claim a resemblance to, or be a semblance of, the most Holy Trinity, it must be something noble, something sublime, indeed the best and most beautiful. The Trinity–sphere idea therefore struck several favorite chords of his system and he could never be tired of repeating it. In this Keplerian understanding and interpretation of the traditional doctrine of the Trinity we can also see a convergence of ideas from several areas:

from religion, from philosophy, from the requirement of rationality, and from geometry.

GOD THE ONE (GOD'S SIMPLE NATURE)

The Christian religion, along with Judaism and Islam, has always upheld that there is only one true God. Theologically this idea is expressed in adherence to monotheism; philosophically this upholds the position that God is a single and simple being, as opposed to a composite one, for a composite being is made up of parts and cannot be perfectly one. There are no parts in God: God is perfect unity. Kepler accepted both of these ideas. He believed that there is only one God and that God's being is perfectly simple.

This religious belief in the oneness and simplicity of God exerted considerable influence on both his philosophical and scientific views. Since for him the universe reflects God, simplicity too has to be reflected in nature. So he argued that nature is simple, and in his scientific study of nature his search would be for theories characterized by simplicity. In fact, he held that the simpler a scientific theory, the more true it would be.

GOD THE RATIONAL

The relationship between religion and rationality has always been controversial. Consideration of rationality in this context has two aspects: rationality insofar as it refers to God's nature and actions, and rationality insofar as it refers to our understanding of the being and actions of God. Is God a rational being? Does God have to abide by canons of rationality? Since God is the source of all rules and laws, should not God be above all laws? Since laws of rationality are the creature of God, to demand God to be bound by them is tantamount to the self-contradiction of a creature being the master of the creator. How can the slave rule over his own master? This quandary has forced many to move toward voluntarism, thereby professing that God's being and actions are not governed by the rules of rationality. Theological voluntarism[32] maintains that the will of God is supreme, and so reason must be subservient to it. Peter Damien, one of the staunchest proponents of voluntarism, argued that because God is omnipotent God can render true even the rationally absurd or contradictory.[33]

At the time of Kepler, theological voluntarism enjoyed a wide and extensive following among intellectuals and common people alike. Kepler's contemporary, the Frenchman Descartes, subscribed to this view. Tycho Brahe also showed strong voluntaristic tendencies. For instance, he believed that his model of the universe was the true one, despite the objection that the Copernican view was more in accordance with the principles of economy and simplicity, because "God the creator manifested his inscrutable wisdom and omnipotence just by his will and power. . . ."[34] Fabricius, Kepler's friend and most regular correspondent, was an ardent advocate of theological voluntarism. Again and again he took Kepler to task for the German's rationalism, which Fabricius thought undermined the supremacy and omnipotence of the Almighty. (The heated debate between the two will be discussed in chapter 3.)

Although Kepler never denied either the omnipotence or the supremacy of God, he believed that God is supremely rational; all the actions of God follow the canons of rationality. He took the divine rationality as a fundamental truth and could not think otherwise. In his writings he repeatedly asserted this claim. For instance, in his letter to Maestlin on August 2, 1592, he wrote: "The creator chose nothing without a plan."[35] Indeed, God, the best and wisest creator, does nothing without sufficient reason. This emphasis on sufficient reason for divine action Leibniz would later make the centerpiece of his own theology.

Kepler's belief that God is supremely rational also influenced both his philosophical and scientific ideas. He believed that nature is rational because, being the reflection of God, nature must be characterized by this divine attribute. This belief was the basis of his lifelong emphasis on the principle of sufficient reason. The significance of this belief for his scientific ideas can be seen from his refutation of Fabricius's voluntarism.[36] If God's will is not governed by any laws, then it will remain unknowable. Since according to voluntarism the phenomena in the world are brought about by the arbitrary will of God, we cannot know the operations of nature, a consequence which naturally forestalls any possibility of scientific knowledge of the working of nature. A voluntaristic God would make science impossible. Thus according to Kepler, the very possibility of the scientific study of nature depends on a rational God. A rational theology becomes a necessary condition for a rational science.

GOD THE GEOMETER

For Kepler God is first and foremost a geometer. This idea was not in the mainstream of Christian thought, although more than three centuries before the birth of Christianity Plato had said, "God is always a geometer." Kepler accepted this view and quoted Plato in his MC.[37] Not something accidental to God, geometry is part of the very essence of God—not something God just has, but something God is. He wrote: "Geometry is coeternal with the mind of God before the creation of things; it is God himself (what is in God that is not God himself?). . . ."[38] The same idea he expressed in MC as well: "The ideas of quantities are and were coeternal with God and God himself and they are still like a pattern in souls made in the image of God (also from his essence). On this matter, the pagan philosophers and the doctors of the church agree."[39] These passages provide reliable evidence for the claim that Kepler believed in a God who is essentially a geometer. (His idea of geometry will be discussed in greater detail in the chapter on his philosophical views.) It suffices to point out here that his God is one who geometrizes, who employed the canons of geometry in the creation of the world.

This view of Kepler had far-reaching consequences. For if God is a geometer and if God imprinted the laws of geometry on the universe at the moment of creation, then geometry becomes the key to understanding nature. This can explain the heavy emphasis he placed on geometry in his scientific method. Geometrical demonstrations and proofs dominated all his major scientific works. Obviously, an integral part of his scientific methodology had its basis in his belief that his God is a geometer.

Geometry had another important impact on Kepler's thought. Since geometrical laws are embedded in nature, understanding nature means understanding geometrical relations. Because these laws are the thoughts of God, understanding nature amounts to thinking the thoughts of God. In this way scientific work through geometrical analysis took on a new significance and meaning for Kepler.

GOD AS ACTIVE FORCE

For Kepler God is an active force. As he put it: "God indeed is substantial energy and subsists in this energy (as I may speak of divine things in a human way)."[40] God is essentially energy, which expresses itself in God's role as creator. According to Kepler, God's creativity has its roots in this essential characteristic.

This concept of God also made significant contributions to the molding of some of his philosophical and scientific ideas. For instance, his dynamic worldview was based on this idea of an active God. As we shall discuss later, the development of the Keplerian idea of force was motivated by his firm belief that God is active force.

GOD THE SOURCE OF LIGHT AND LIFE

The idea that God is the source of light and life is an old one that occurs very clearly in both the New and the Old Testaments. The Psalmist says: "The Lord is my light and my salvation" (Ps. 27:1). Again, he proclaims: "For with Thee is the fountain of life..." (Ps. 36:9). The New Testament, especially the Gospel of St. John, has many direct references to this theme. The Bible not only considers God the source of light and of life, but even identifes God with them.[41]

Kepler also accepted this idea and often wrote about it. He spoke of Christ as "the Son of God,...that true light that enlightens every person coming into this world...."[42] According to him, the divine Being emits some kind of radiation. For instance, he wrote in HM: "The souls of Christians are copies of the creator and are sustained as such by him through a certain 'irradiation' of the divine countenance into them."[43] On another occasion, writing to Johann Georg Brengger about the creation of spiritual beings, Kepler brought up the same theme.[44] In his view in all creatures, even in spiritual ones, were two things: one like matter, which he thought was the most subtle aethereal body; and the other like form, which he thought was the ray of the divine face. God was the source of a special radiation.

In developing these ideas Kepler was influenced not only by the Bible but also by the Neoplatonists like Plotinus and Proclus, who identified the supreme being as the One. From their writings it is also very clear that they often referred to the One as the Light.[45]

Closely associated with the idea of God as the source of light is the theme of emanation from God. Emanation theorists like Plotinus and Proclus[46] argued that just as light came from God, a certain emanation also had its source in God. According to these Neoplatonists, the whole creation was an emanation from God.

These ideas of light and emanation had a crucial influence on Kepler. As we shall see later, Kepler believed that the sun was the visible representation of the invisible God the Father. The close relationship between God and light must have played a decisive role in the origin of this important belief.

GOD THE MUSICIAN

The musician God, the lover of cosmic harmony, was another characteristically Keplerian idea of the Almighty. God's actions, according to him, are governed not only by the laws of geometry but also by those of harmony. In his view, geometry and harmony complement each other, and we need to pay attention to both for a complete knowledge of the secrets of nature. (This aspect of harmony will also be discussed in detail in the chapter on Kepler's philosophical ideas.)

The idea of God as musician or lover of harmony was a clear example of the influence of empirical inquiry on Kepler's religious outlook. In the MC he emphasized the supreme importance of geometry. Very little was said about harmony, especially musical harmony. In the initial stages he was very much convinced that geometry alone was sufficient to unlock the cosmographic mystery. However, as we shall see in chapter 2, when the geometrical theory of the structure of the solar system was subjected to empirical (observational) test, he realized that geometry alone could not account for the laws of nature. He realized that God not only geometrized but also harmonized. This was the major theme of the HM. Hence empirical investigation led him to modify his idea of God.

THE KEPLERIAN IDEA OF HUMANKIND

The Place of Humans in Kepler's System

Kepler's idea of humankind was intimately associated with his idea of God. His idea of humankind, just like his idea of God, combined traditional and specifically Keplerian elements. Here, too, we can see the transitional character of his works and ideas. He considered the human being as the image of God. As we know, this idea is at least as old as the first chapter of Genesis. Although he accepted the traditional idea, he gave it a new interpretation.

Despite its antiquity and constant use by scholars and nonscholars alike, this metaphor has not met with any unanimity of meaning and interpretation. According to the *Dictionary of Biblical Theology*, the metaphor, "the image of God," conveys the unparalleled dignity of the human being in creation. In the Jewish tradition, God is considered so transcendent that it is forbidden to make any image of God. This prohibition reveals how lofty is the majesty of the Almighty. Since the human being is the image of God, this metaphor shows the unique dignity of human being.

As to what this likeness consists in, a variety of opinions have been expressed. Some scholars point out that the likeness is based on the ability to exercise power over earthly creatures: God has absolute power in this respect, and humans also have a certain amount of power; hence there is a likeness between God and us. Some others hold that the basis of the likeness lies in the ability to create; God is the creator of living beings, while humans can be the procreator of living beings. Still others present immortality as the basis of resemblance, stressing that God is immortal and the human being is endowed with an immortal soul. Some other scholars believe that the likeness is to be sought in the spiritual qualities of humans—our capacity of self-consciousness and self-determination—for God has a distinct personality and so do we. Traditionally, very often rationality has been pointed out as the basis of likeness; God is supremely rational and among all the visible creatures only we have the quality of rationality, so in this way the human is the image of God.

The interpretations given above are based on particular scriptural passages or on certain philosophical positions. Kepler subscribed to many of these traditional positions but went beyond them. For him, the human being is the image of God precisely because the characteristics of God are reflected in the human being. Not just a casual similarity of some accidental attributes, Kepler's idea goes far beyond into the realm of essence. As he put it: "Human souls are created in the image of God the creator, conforming to him in essentials as well."[47] Now let us discuss some of the divine characteristics Kepler saw reflected in humans.

THE HUMAN BEING AS CO-CREATOR

Just as God is the *optimus creator*, the human being is a co-creator in a certain sense. Although creation out of nothing is beyond our power, we have the ability to procreate our own species and make innumerable new artifices. I have indicated that God's role as creator flows directly from God's essence as active force. Kepler could see a similar relationship between human activity and human energy: the fact that humanity is active or energetic entails procreativity.

THE HUMAN BEING AS RATIONAL

One of the main reasons why Kepler considered the human being the image of God was rationality, because human rationality he considered a mirror of divine rationality. He believed that humankind is rational and human actions are governed by rational considerations. Kepler argued that although humans, unlike God, are not omniscient, in certain limited cases human knowledge is as perfect as divine. Thus, just as God is rational and is capable of knowing, we, too, are endowed with these capabilities.

THE HUMAN BEING AS GEOMETER

The distinctly Keplerian interpretation of the human being as the image of God was based on the unique importance he attached to geometry. God for him is most of all the geometer who thinks and acts primarily according to the canons of geometry. Similarly, geometry is the foundation of the likeness or imageness between God and the human being. "Geometry is God himself. . . . With the image of God it has passed into man."[48] "What is in the mind of man except numbers and quantities?"[49] Just as the content of the mind of God at least in part is geometry, so too the human mind contains the same principles and laws. Geometrical laws "are within the grasp of human understanding. God wishes us to know these since he has furnished us in his own image, that we may come into participation of the same reasoning with himself."[50] At creation God infused geometrical laws into our minds; consequently, we are capable of thinking the thoughts of the creator. This, according to Kepler, is one of the most important reasons why it is said that God created the human being in God's own image. In this way, he interpreted one of the fundamental claims of the Christian religion in a unique manner, basing himself on his own special idea of God.

Kepler's position has a crucial epistemological consequence because geometry takes on an unrivaled significance in human knowledge. Geometry is universal in its accessibility, since all humans have it as part of their essence, and is universal in its application, since it has become the universal means of acquiring knowledge. This branch of mathematics can help us acquire knowledge about God because God is a geometer. It can reveal human actions because the human mind operates in accordance with its laws. It can also reveal the secret laws of nature, because nature has been created according to geometrical laws and principles.

Another significant epistemological consequence of this position was that it convinced Kepler that not all the hunches and intuitions that arose in the human mind could be false. In fact, many of them must be true. He wrote to Herwart: "Not every hunch is wrong. For man is the image of God, and it is quite possible that in regard to certain things that make the ornament of the universe, his opinion is the same as God's. . . ."[51] Since the human mind operates in accordance with geometrical laws, such hunches cannot not go amiss and must lead us along the right path. He seemed to have put this optimism into practice often in his research on the planet Mars. Thus, although often to others his work seemed to go nowhere, he could remain optimistic because he had a hunch that he was on the right track.

A serious objection to Kepler's position discussed above is that it eliminates the traditional distinction between God and humanity, between the infinite and the finite. If God and humanity are essentially the same, then what is the difference between them? Also, he seemed to undermine the age-old difference between divine and human knowledge. Traditionally, a wide gap exists between the two, but he said: numbers and quantities "do we rightly understand and if it can be said with reverence, with the same kind of thought as God, as much indeed as we do comprehend these things in our mortality [i.e., limited nature]."[52] Concerning geometrical knowledge, the difference between God and humans is only quantitative, not qualitative. God knows an infinite number of truths whereas humans know only a finite number; but with regard to such mathematical truths as the theorem that parallel lines never meet, our knowledge is as perfect as God's. To most of his contemporaries this was heresy and presumption in its worst form, clearly a case of a man playing Lucifer to make himself equal to God. To his critics Kepler retorted: "Foolishly do they fear that we shall make man God, the counsels of God are inscrutable, not his corporeal works. For what are the works of God, if they are compared with his counsels? The counsels [consilia] of God are God himself, but the works are his creatures, and it is not a great thing for God to create man capable of understanding his works."[53] According to Kepler, his opponents' fear arose because they failed to pay attention to the twofold distinction between the counsels of God and the works of God, and between what is quantitively the same and what is qualitatively the same. The perfect divine-like knowledge referred only to the works of God, not to God's plans or counsels. The counsels of God still remained inscrutable to us humans. Besides, the quantitative difference between divine and human knowledge he was always ready to admit. A clear distinction between the two still existed.[54]

THE HUMAN BEING AS DYNAMIC BEING

Kepler argued not only that God is substantial energy but also that humans share in this substantial energy. According to him, this was yet another compelling reason to believe that the human being is the image of God because "the essence of the divine image consists in this energy. . . ."[55] Like God, humans are also a source of energy and force and so we are the image of God.

THE HUMAN BEING AS RECIPIENT OF GOD'S LIGHT

Just as God is the source of light or radiation, the human being is the recipient of that light. He believed that the irradiation from the divine countenance was necessary for the preservation of human beings.[56] In his letter to Brengger in 1608 he explained the creation of humans as follows:[57] just as the ray of the sun makes green color correspond to green objects, red to red ones, etc., so also the ray of the divine face falling on properly prepared matter and spirit makes them become like the divine by infusing them with life and rationality. Humans are humans because they have received the divine ray or radiation. The falling of this radiation makes humans the image of God because it makes them become like God.

THE HUMAN BEING AS LOVER OF HARMONY

Another aspect where Kepler saw the reflection of God in humanity was in considerations of harmony, especially musical harmony. God loves harmony and so do we. In his opinion the human soul has the inborn ability to detect and appreciate harmony. For instance, he quoted the example of illiterate farmers who can detect melodious, well-proportioned musical beats and dance to them. He believed that such a phenomenon could arise only because humans are the image of God the musician and the lover of harmony.

Here Kepler reveals a striking parallel between God and the human being. However, the parallelism does not threaten the age-old distinction between God and humans. For one thing, not all the characteristics of God are reflected in us. For instance, we do not have any real parallel to God's essential trinitarian nature. Again, God and humans do not share characteristics in the same way: what one sees in humans only reflects, not perfectly duplicates, and even this only in limited items. There is no danger of humans becoming equal to God.

THE KEPLERIAN IDEA OF NATURE

The Place of Nature in Kepler's System

A view of nature, the universe, was an integral part of Kepler's religion. In contrast to the dominant medieval outlook, the world for him was not just a rest station for human wayfarers on their journey to their heavenly home. He held extremely sacred and positive beliefs about the universe, the "bright temple of God."[58] Involvement in the world was not a burden imposed on the human race in the Garden of Eden: the universe and involvement in it through scientific research was his sure means to reach heaven.[59] As he wrote to Fabricius, nature aspires to divinity.[60] The positive ideas he had toward the created world rendered his scientific work meaningful and purposeful.

From a philosophical point of view as well, tradition held that the world was imperfect and we could not have a perfect knowledge of nature. Plato had argued that the Demiurge could not produce a perfect world but produced the various things in the universe by imposing copies of eternal Forms onto matter. Since the Forms did not fully fit with matter, however, the material things produced by this process remained imperfect. Not only that, but Plato also believed that matter had a tendency to wiggle out of the influence of the infused or imposed Forms. Thus, according to the view expressed in his *Timaeus*, the material world was neither perfect, nor could we have a perfect knowledge of it. Aristotle also believed that we could not have perfect knowledge of the physical world because knowledge about it could not yield perfect accuracy. Hence both metaphysically and epistemologically the world was imperfect: metaphysically because it had been produced imperfectly; epistemologically because it was never amenable to accurate scrutiny and investigation.

In contrast, Kepler had a positive and optimistic view toward the world. The universe, he believed, is the manifestation of God. He expressed his basic thoughts on the universe in a letter to Herwart written on April 10, 1599: "The world is the corporeal image of God, whereas the soul is the incorporeal, though created, image of God."[61] The material world is the image of God become tangible, taken concrete shape, while the world of spiritual beings is the same image of God in incorporeal form. Furthermore, God ordained that the universe act and operate by the same laws as God's: "As God the creator played, so he also taught nature, as his image, to play; and to play the very same game he played for her first. . . ."[62] Since these claims have far-reaching consequences, they deserve very careful scrutiny.

While discussing Kepler's idea of God, I identified a number of items, all of which can be divided into two categories: (1) those referring to the inner nature or makeup or structure of God, to speak in human terms, since, strictly speaking, such a language is not applicable to God who is perfect simplicity; (2) those referring to the attributes of God. In the first category one can include "God the one," "God the Trinity," and perhaps, "God the *optimus creator*." The other items come under the second category. Just as Kepler could see these items reflected in human beings, he could see a similar mirroring in our universe. Thus he saw a reflection of the divine

Persons in nature both with regard to the interrelationship among the three Persons in the Trinity and, to some extent, in their functions. Moreover, he saw the different attributes of God in a noticeable way reflected in the material universe.

Reflection of the Divine Persons in the Material Universe

REFLECTION OF THE TRINITARIAN "STRUCTURE" IN THE MATERIAL WORLD

Kepler believed in the Holy Trinity and could find meaning in it because he was convinced that this mystery was reflected in the structure of the universe. Our spherical universe is an image of the triune God: "The sphere possesses a threefold quality: surface, central point, intervening space. The same is true of the motionless universe: the fixed stars, the sun, and the aura or intermediate aether; and it is also true of the Trinity: the Father, Son, and Holy Ghost."[63] This is about the motionless universe. A similar relationship exists in the case of the mobile universe as well; it is made up of the sun and the known planets of the day. "The sun in the middle of the moveable, being immoveable itself and yet the source of motion, bears the image of God the Father, the creator. Now what creation is to God, so is motion to the sun. Thus it moves [the bodies in the space within] the fixed stars, just as the Father creates in the Son. . . . Again the sun disperses a moving power through the medium in which are the moveables, and in just this way the Father creates through the Spirit or through the power of the Spirit."[64] Thus the general structure of the universe is modeled after the trinitarian God. Just as the Trinity has three distinct, yet intimately related, parts and none can be complete without the other, the universe, too, has three parts that are intrinsically interrelated.

Is the similarity just heuristic, just an analogy designed to make the mystery of the Trinity intelligible to simple people? Kepler certainly wanted to make the mystery as intelligible as possible, but the trinitarian model, far from being a figment of his imagination for heuristic purposes, is an archetype of the universe, a real blueprint of the universe. In his own words, "Nor should it be taken as a meaningless resemblance, but it should be reckoned as one of the causes, as a form and archetype of the universe."[65] According to him, the universe does literally have a trinitarian structure.

Kepler saw God's creative work mirrored in the universe. He compared the creative power of the Father to the motive power of the sun. The "outflow" of the Father gave rise to created beings; the "outflow" of the sun gives rise to planetary motion. The Father carried out his creative work through the Holy Spirit; in the same way, the sun diffuses and expends its motive power through the *intermedium*. In another place he drew a parallel between the first beginnings of creation and the eternal generation of the Son: "There follows, then, the straight line, which by the movement of a point located in the center [of the sphere] to a single point on the surface, represents the first beginnings of creation, emulating the eternal generation of the Son in that the center flows out towards infinitely many points of the whole surface, which, under the rule of the most perfect equality, is formed and described by infinitely many lines. . . ."[66]

It should come as no surprise that his comparing the universe to the Trinity is considered bizarre by present-day Kepler scholars, just as it was by his contemporaries. Not a single one of his critics has been highly impressed with such a scheme. Even the often-sympathetic Koyré considers the idea odd and absurd, although he does point out that this strange view produced positive and beneficial results in Kepler's scientific research. One might argue that the astronomer saw the reflection of the Trinity in the universe simply because he was prejudiced in favor of his religion. He was looking through his own colored glasses and saw what he wanted to see; someone else with a different background would have come up with a totally different conclusion. To be sure, his arguments in this context cannot be truly conclusive evidence for anything. Nevertheless, the crucial question to be asked is this: Did he see what he saw because he wanted to see it, or because it formed an integral part of his larger system of thought, in terms of which it could be understood as quite reasonable to expect such a conclusion? If he had good reasons to expect it, then his conclusion deserves respect. When a person sees what he or she wants to see even though there is no good reason to expect it, when it is not a consistent part of a larger coherent view, then we may appropriately label the conclusion prejudicial and unworthy of serious consideration. A person who has been cheated by strangers many times in the past cannot be accused of being unreasonable if she is suspicious of the next stranger she meets, although this time the stranger is not a cheat. For she has good reasons to expect that she may be cheated again. Kepler seems to have had good reasons to expect that the world was created by God after the model of God's own nature. If God found the trinitarian "structure" best for God, then it was only natural that the best and wisest creator who wanted to create the best universe should choose a trinitarian structure for creation. Similarly, geometry is the essence of God,[67] if the sphere is the most perfect geometrical structure, then it should not be surprising that the Trinity is spherical in structure.

The First Person of the Trinity:
The Sun as the Reflection of God the Father

That the sun had a unique place in the thoughts and works of Copernicus is quite evident from the memorable passage he wrote in DR:

> At rest, however, in the middle of everything is the sun. For in this most beautiful temple, who would place this lamp in another or better position than that from which it can light up the whole thing at the same time? For, the sun is not inappropriately called by some people the lantern of the universe, its mind by others, and its ruler by still others. [Hermes] the Thrice Greatest labels it a visible god, and Sophocles' Electra, the all-seeing. Thus indeed, as though seated on a royal throne, the sun governs the family of planets revolving around it.[68]

The metaphors used here evoke the sun's royalty and divinity. His wholehearted acceptance of this Copernican notion of the sun Kepler makes clear from his own writings. For example, in the *Dissertatio cum Nuncio Sidereo* he wrote: "The sun certainly is in the center of the universe, it is the source of light and heat, it is the origin of life and movement of the world."[69] In the MC he fully endorsed the many royal and divine epithets ascribed to the sun.[70] All these passages express that for him the sun occupied the most central position in the universe, exercised royal authority over all other creatures, and even resembled God.

The divine aspect of the sun was brought out even more clearly when he described the heavenly court composed of the sun surrounded by the planets. Then he added that the planets were remaining "veluti adorantium" (as it were adoring the sun).[71] This picture evoked the godly status of the sun.

God is often spoken of as the Alpha and the Omega, the beginning and the end of everything. A similar idea was developed about the sun as well. The sun was the starting point and the terminal point of light. "Not only does light go out from the sun into the whole world, as from the focus or eye of the world, as life and heat from the heart, as every movement from the king and mover, but conversely also by royal law these returns, so to speak, of every lively harmony are collected in the sun from every province in the world. . . ."[72]

That Kepler saw God in the sun with all these unique characteristics was evident from his comments on the hymn of Proclus, whose views he endorsed for the most part. Just as Proclus while invoking the sun was invoking the God who created it, Kepler too saw God the creator when he contemplated the greatness of the sun.[73] This, of course, should be only natural: just as God is at the center of the heavenly kingdom, the sun is at the center of the material kingdom. Just as at creation God set everything in motion, the sun sets and keeps the planetary bodies in motion. God is the source of life and light, the sun is the source of life and light in the material world. Theology claims that the operations of God cannot be explained in terms of ordinary physical capabilities. Similarly the sun's rotation could not be explained, according to Kepler, by any ordinary mechanism: it needed a special vital capability.[74] Indeed, his reasons to draw a parallel seemed to be quite clear.

In the visible sun Kepler was seeing the representation of the invisible God. A clear perception of this point brings home how completely off the mark are those critics who accuse Kepler of sun worship, obsession with astrology, and the like. Such accusations betray a lack of understanding and appreciation for what he was up to. He was too clever and too religious-minded not to have noted the difference between creator and creature, not to have known who alone deserved worship. Indeed, while talking about the sun, he did use words reserved to the supreme Deity; but in addressing them to the sun, he was, in fact, worshiping the One who is the creator of the sun. The sun occupied a unique place in Kepler's thought and works because it was for him the visible image of his invisible God.

The Third Person of the Trinity:
The Active Space between the Sun and the Fixed Stars
as the Reflection of the Holy Spirit

Kepler saw a parallel between the Holy Spirit and the solar–emanation–filled *intermedium*. The Holy Spirit is ever active and ever dynamic, so also is the emanation-filled space between the sun and the planets. The Father acts through the Holy Spirit, the sun also acts on the planets through this *intermedium*.[75] The solar emanation can act upon the planets and change their position, just as the Holy Spirit can act upon and cause changes in humans. All these ideas convinced him of the similarity between the Holy Spirit and the solar–emanation–filled space. As we shall see later, these considerations helped Kepler shape his idea of force. Hence his idea of force had its foundation on this religious position.

The Second Person of the Trinity:
The Mirroring of the Mystery of the Incarnation
in the Material World

In the mystery of the Incarnation, another parallel between what happens in the heavenly kingdom and what happens in the material world was found, though in a way much less evident compared to the others we have discussed. Earlier I pointed out the parallel between God the Son and the outer surface of the cosmic sphere; there the comparison functioned with respect to the "structure" of the two worlds. Here the parallel is founded on the happenings in the two worlds. The basic idea of the mystery of the Incarnation is that God the Son came down to the world and became man without losing his divinity, in order to redeem and elevate the fallen humankind; this was willed by the Father as a supreme act of love towards humankind. It seems that Kepler saw a similar event with regard to the position of the earth, the abode of humankind. Copernicanism had unseated the earth from its privileged place, but the earth had been able to recover a privileged position, though not of the same dignity as the pre-Copernican one, just as redeemed humankind's status was not the same as before the Fall. This parallel to the Incarnation had come about because God the creator ordained that just as God the Son lowered himself to the level of a human being, so also the sun should be lowered to the status of a planet so that the earth would have three bodies inside, and three outside, of its orbit. Kepler himself described the situation:

> Therefore in my opinion it was deemed by God fitting for the Earth . . . to go round in the midst of the planets in such a way that it would have the same number of them within the embrace of its orbit as outside it. To achieve that, God added the sun to the other five stars [i.e., planets], although it was totally different in kind. And that seems all the more appropriate because, the sun above being the image of God the Father, we may believe that by this association with the other stars it was bound to provide evidence for the future tenant of the loving kindness and sympathy which God was to practice towards men, even as far as bringing himself down into their intimate friendship.[76]

This parallel has a number of problems. First of all, Kepler confuses the picture by bringing in God's involvement with humankind in the Old Testament. Again, in the Incarnation, God the Son lowered himself, but in the material world, the sun, the representation of the Father, "lowers" itself.

Nevertheless, several characteristics of Kepler's explanation of the special position of the earth convey such a striking similarity to the Incarnation that I am inclined to believe that he did intend the parallel. For instance, in both events occurs a lowering to an inferior level without the suffering of any loss of dignity. The Incarnation was motivated by God's deep love for humans, and in Kepler's account the sun's lowering of its position was also a mark of love. The Incarnation elevated the dignity of humankind; the arrangement of the bodies in the solar system elevates the dignity of the earth.

REFLECTION OF THE DIVINE ATTRIBUTES IN NATURE

Nature as Ordered Unity

According to Kepler, the universe is characterized by an ordered unity because "God, like one of our own architects, approached the task of constructing the universe with order and pattern, and laid out the individual parts accordingly. . . ."[77] This is to be expected, since, being a reflection of God, the universe or nature should reflect the oneness and the trinitarian nature of God. Kepler saw this ordered unity beautifully expressed in the ordering and arrangement of the heavenly bodies as given by the Copernican model.[78] Insofar as oneness can be looked on as absolutely perfect unity, a unified universe reflects, although very imperfectly, God's oneness. Insofar as the trinitarian nature and life of God can be seen as absolutely ordered existence, an ordered universe may be considered a faint reflection of the trinitarian God. The order and unity that Kepler saw in the universe had their basis in his idea of God.

Nature as Rational

Kepler believed that nature is rational; nature does nothing without sufficient reason. This obviously is also to be expected if the universe is a reflection of a supremely rational God.

Nature as Mathematical (Geometrical)

According to Kepler, the universe is geometrical in the sense that it was created in accordance with the laws of geometry. When God created the universe, God used the geometrical archetypes: they provided the creator with the blueprints for the fashioning of the world. The universe has to be geometrical because it is the reflection of the geometer God.

Nature as Dynamic

In Kepler's view the universe is dynamic, ever active, not a static structure but an ever-energetic machine performing innumerable complex and complicated activities. Since God is active force or energy, nature proves here, too, to be a true reflection of the creator.

Nature as Harmonious

Harmony is another characteristic feature of the Keplerian universe. Already in MC he believed that harmony pervaded our universe. Harmony in this context meant the deducibility of everyhing from a single, fundamental postulate.[79] Thus he believed that the Copernican system was harmonious and true because it could explain various phenomena by means of a single postulate of a moving earth. Not peculiar to Kepler alone, Copernicus also valued this idea of harmony immensely. Kepler believed that his theory was true because "we find underlying this ordination (arrangement) an admirable symmetry in the universe and a clear bond of harmony in the motion and magnitude of the spheres such as can be discovered in no other wise."[80] In this harmony of the world one can also see a reflection of God. Although there are three distinct persons in the triune God, they constitute one, true God; the three different Persons are united by absolutely perfect harmony.

Consequences of Kepler's Idea of Nature

Kepler's view that the universe is the image of God become concrete or tangible had significant consequences. Before we discuss them, a few comments are in order. That the world was created in the image of God was no accident. God who does nothing by chance had good reasons for doing so. Once having decided to make the best and most perfect world, God had to look for the best blueprint possible. Nothing else could have been chosen, not even an angelic model. In Kepler's own words: "It will be absurd for God to create the universe, the most excellent being, in the image of the angels rather than of God."[81] Even an angelic form would have fallen short of God's plan to create the best and most beautiful world.

Most Kepler commentators seem to hold that his geometrical ideas were the sole basis for his claim that the universe is the image of God. For example, Caspar explains this point as follows: the imageness arises because of the special characteristic of quantities, which consists in their being able to form relationships, since they allow comparison among themselves.[82] I have argued that Kepler's arguments are far more broadly based. Geometry provides only some among the very many ideas required for drawing his conclusions.

Another observation has to do with the image metaphor. Earlier I indicated that for Kepler the human creature is the image of God, but here he holds that the universe is the image of God. Does this mean that both humanity and the universe are the image of God in the same way? Despite an ambiguity in the use of the image metaphor, a clear distinction is to be found between the imageness of humanity and that of the universe. Consider how geometry becomes one of the bases for the imageness: the human being is the image of God because, just as for God, the human mind also contains geometrical principles; just like God, humans think in terms of geometrical laws and act in accordance with them. On the other hand, the universe is the image of God because it has been created according to the laws of geometry; geometrical in structure, the universe has geometrical relationships and properties that are embedded in its various creatures and components. Humans are the image of God because as knowers we are characterized by geometry; nature or the universe is the image of God because as known and knowable it is characterized by geometry. For humans and the universe, geometry is the basis for the likeness, but in two different ways: in the humans by virtue of our being active knowers, of having the capacity to know; and in the universe by virtue of its being a passive known, of having the capacity to be known.

PARALLELS BETWEEN THE DIVINE AND MATERIAL WORLDS

Kepler constantly saw a parallel between the supernatural and the natural, between the divine and the material. He looked for further instances of this parallel in the universe. For him the material world reflects the heavenly or divine world. This situation is analogous to the fundamental Platonic postulate dividing the universe into the worlds of Ideas and of material things, the latter being a reflection of the former. However, unlike Plato, Kepler believed that the material world, though a reflection of the divine, is perfectly real. I suggest that this conviction and this search guided Kepler in his long Odyssey to the discovery of the laws of planetary motion.

POSITIVE VIEW OF NATURE

The traditional Christian view of the material world has tended towards the negative, especially in Kepler's time. In contrast, Kepler had an extremely positive view since for him the material world is really and truly the image of God become manifest in a concrete way. At the same time, he never allowed his ideas to degenerate into any form of pantheism because the material universe, never identified as precisely equivalent with the Godhead, is only a reflection or image—and an imperfect and faint image at that.

That Kepler strongly opposed a negative view of the world can be seen from his scathing and almost merciless attack on Pistorius,[83] who was a representative of the traditional view. In one of his letters to Kepler he noted that he was seriously ill, that he could feel the end fast approaching, yet he was of good cheer because he knew that soon he would be freed of the vanities of the world and would be led to his true heavenly home to participate in his heavenly heritage.[84] For the old man, the world offers little that is precious or valuable; it is something to be gotten rid of, the sooner, the better, and the more one renounces the world, the better off one is.

Uncharacteristically, Kepler wrote a merciless and sharply worded response to his friend Pistorius. He did not try logically to refute his friend's views, rather he unleashed a passionate and sarcastic *ad hominem* attack against Pistorius's Catholic religion. The gist of his reply was that the emphasis on the "inanities of the world" and the disdain for the present world, instead of being a source of sanctity, were, in fact, at the root of abuses and corrupt practices in the Catholic church. He wrote: "Certainly I do not doubt that you are prepared by your trust in Christ the Savior, and hope of a share in his celestial inheritance, and by your contempt, as you write, and, hate and regret, as I interpret, of the inanities of the world."[85] Then in a long parenthesis he added that from this hatred for the "inanities of this world" stemmed the passionate factional fights, the views about personal beatitude, the Roman claims about religious supremacy, the abuse of power, etc.

Kepler scholars do not seem to find an adequate reason for this unusual reaction. Caspar and others think that the unhealed wounds of past bitter experiences of religious persecutions by Catholic rulers excited him to break forth almost uncontrollably.[86] There is some plausibility for this explanation, since Pistorius was a convert from the Reformed creed and he had been very active in polemics. However, these things were well known for a long time, and so were nothing new to Kepler. There is no good reason to believe that they alone could have extracted such a reaction from him. I think that the most potent reason for his unusual outburst was his friend's condemnation of this world as something worthless. For Kepler the world is too sacred to be counted as "inanities."

POSITIVE VIEW TOWARDS THE HUMAN BODY

In the traditional Christian view only the soul was considered to be the image of God. The body was looked upon as something inferior;

in fact, it was often considered as the source of sin.[87] Mortifying it and bringing it under the submission of the soul were considered necessary for any growth in sanctity. The body thus was looked down upon. But Kepler's view ennobled it: The body, too, is the image of God, it is the corporeal image of God. The body and soul are two aspects of the same person, both noble and sublime.

THE TRADITIONAL DISTINCTION OF BODY-SOUL, MATTER-SPIRIT CHALLENGED

According to Kepler's position, the traditional distinction between body and soul, matter and spirit, seems to evaporate. The world is the corporeal image of God whereas the soul is the incorporeal image. Both are intimately linked, both find their roots in God, both mirror God. In fact, one can say matter and spirit are just two aspects of the same reality.

SACRED CHARACTER OF THE STUDY OF NATURE

God's self-revelations flow not only from words, but also from deeds. This point is at the basis of the frequently recurring theme of the "Book of Nature." According to Kepler, nature is a sacred book with a sublime message to all humankind. As he put it in the *Epitome*, "This is the very Book of Nature in which God the creator has proclaimed and depicted his essence and his will toward man in part and in a certain wordless kind of writing."[88] Just as we can come to know the personality and greatness of an author through his or her book, we can come to know God through the Book of Nature. In fact, the self-unfolding God "wishes to be known through this Book of Nature."[89]

The specialty of Kepler's interpretation of this theme consisted in relating the Book of Nature to the Book of Scripture in an original way. He placed them on a par. Both are aspects of one and the same reality, complementing each other and thereby giving humans a further and more complete manifestation of God. He argued that since God has mouth and hands, God reveals through both, the word of God proceeding from mouth and the deed of God from hands. The Book of Scripture recounts the word of God, whereas the Book of Nature the deed of God. Hence both books are sacred, both are worthy of our total respect and attention. This conclusion has an extremely significant consequence: science, which is the study of the Book of Nature, becomes a profession very closely analogous to scriptural theology. Thus this theme is at the basis of his perception of the nobility of science. Obviously, his belief about the nobility and importance of science follow from his religious ideas. The theme of the Book of Nature was neither new nor unique to Kepler, since some of his predecessors[90] and contemporaries talked about it. For instance, Galileo spoke of the Book of Nature written in the language of mathematics. What was special about Kepler was the distinctive way he used the theme and the unique role it played in forming his mode of thinking and acting. This idea in a way revolutionized his whole life, providing it with a new direction. To be a priest of God in the Lutheran church was his great ambition. Accepting the job of a mathematician-astronomer in Graz he viewed only as a temporary avocation. But the perception of the full significance of the Book of Nature transformed his vocation. He now realized that he could be both a priest of God and an astronomer.

Sacred Character of Astronomy

It was obvious to Kepler that if the Book of Nature is something sacred, then the study of nature has to be something sublime, just like the study of Scripture. For a person to whom the universe is the "sacred temple of God" this conclusion has to follow, since astronomy is nothing but the study of this sacred temple. But the theme of the Book of Nature elevates astronomy to an even higher plane. Astronomy is not just the study of the temple or abode of God; it becomes the study of God manifested in and through nature, just as the study of the Book of Scripture is not just the study of God's verbal communication but also of God communicating to us. He affirmed this sacredness again and again in his writings. For instance, in AN he asserted that it was "the divine voice that calls humans to learn astronomy."[91]

That the study of astronomy is something noble and sublime also was no new idea. After all, already the Pythagoreans, Plato, and the other ancient Greeks believed that the heavens were the abode of the gods and any study of them could be considered a pious act. An epigram in some of the manuscripts of Ptolemy's treatises confirms this belief: "I know that I am mortal by nature, and ephemeral: but when I trace at my pleasure the windings to and from of the heavenly bodies I no longer touch earth with my feet: I stand in the presence of Zeus himself and take my fill of Ambrosia, food for the gods."[92] Copernicus expressed a similar view when he referred to astronomy as "a divine rather than human science, which investigates the loftiest subjects. . . ."[93] Thus had Pythagoras, Plato, Ptolemy, Copernicus, and others recognized the nobility of astronomy. Kepler's conclusion was nothing new: but his reasons were. For him astronomy was not a study of something incidental and external to God but was a study of something intimate to God.

Astronomers Are Priests

One of the direct consequences of placing the Book of Nature and the Book of Scripture on a par with each other was that, for Kepler, astronomers became priests of the Almighty. Just as ordinary priests are ministers of the word of God, astronomers are ministers of the deed of God. Ordinary priests give glory to God by expounding the mysteries in the Book of Scripture, whereas astronomers do the very same by explaining the mysteries in the Book of Nature. He emphasized this conviction repeatedly in his correspondence with friends: "Indeed I am of the opinion that since astronomers are priests of Almighty God with respect to the Book of Nature, we should concern ourselves not with the praise of our cleverness but with the glory of God."[94] This was no merely pious statement, as far as he was concerned. Nor was it offered as a rationalization to justify to himself and to his relatives and friends his decision to discontinue his pursuit to become a Lutheran priest. He really meant what he said, as was evident in the way he lived out his conviction: with the zeal of a priest-missionary, he fully dedicated himself to astronomy with utter selflessness. Even in his most mature age he remained faithful to his conviction, as could be seen from the fact that he wanted the *Epitome* to be interpreted as a hymn that he composed as the "priest of God at the Book of Nature."[95] His long reluctance to compose Tycho Brahe's refutation of Ursus, too, could be understood, at least in part, from this perspective. For, according to Kepler, as an astronomer Tycho must be concerned about the glory of God, not about the gratification of his ego and false pride. It was common knowledge that the principal reason for the Dane's insistence on a refutation even after the death of his victim was to boost his own ego.

The belief that astronomers are priests of God the Almighty had a crucial influence on Kepler's thoughts and works. No one can fully follow his later works and train of thought without fully understanding and appreciating the importance he placed on this conviction. I have already pointed out that in all his major works he had only one goal in view: to praise and glorify God as God's priest. This kept him going even when the road was filled with formidable obstacles. This kept him from giving in to disappointments over many failures, because the priest of God does not look for his own comforts and glory, he is happy as long as God is glorified in his efforts.

Furthermore, this conviction defined the goal of science or astronomy for him. The aim of science was to discover the plan of God, the thought of God, not to play God over nature, not to have power over nature so as to control it, as Francis Bacon would have it. It was to discover in this plan God's great wisdom and love for humankind so that we can praise the Divine Majesty all the more.

SCIENCE AND RELIGION RECONCILED

The theme of the Book of Nature and his consequent belief that astronomers are priests of God led Kepler to believe that science and religion collaborate rather than contend. From this theme it followed that both ordinary priests and astronomers have a sublime vocation to perform a sacred function, and their works complement rather than compete with each other. One can be both a scientist and a believer, for between science and religion is no real conflict.

This positive view about the interrelationship between science and religion can explain his emphasis on the importance of explanations in terms of physical principles and forces. Most of his contemporaries and predecessors believed that interpreting natural phenomena by means of physical causes engendered atheism and undermined belief in God because it attempted to exclude God. Fabricius, for instance, subscribed to this view. However, Kepler argued that such fears were unfounded and arose from a lack of right perspective. Given his view of the universe, it becomes clear that physical principles and forces are the manifestation of God and using them redounds to God's glory rather than reduces it. Therefore scientific explanation can and does add to our knowledge of God's glory rather than detract from it. No robber or enemy of God, science is God's devoted servant and admirer. In this way Kepler could experience no tension between being a faithful believer and a dedicated scientist.

KEPLERIAN IDEA OF RATIONALITY IN RELIGION

Rationality as a
Characteristic Mark of Religion

Rationality was the fourth element of Kepler's religion. We have already discussed God as supremely rational, humanity as sharing in this rationality, and nature manifesting rationality in its structure and operations. Here our main concern is with our understanding of God and God's actions. Can we have a rational understanding of divine nature, operations, and revelation in the world? Since religion is centered around God's being and revelation, this discussion is directed especially to religious beliefs. Can religious beliefs and claims be understood rationally? Or are they simply irrational or suprarational? Can a religious person be a faithful believer and a rational person at the same time? If the answers to these and related questions are in the affirmative, then what role does reason play in religion? If the answers are in the negative, how can religion defend itself against consequent accusations of emotionalism, unpredictability, and inconsistency?

The traditional answers to the questions above have been neither simple nor straightforward. Some Christian scholars believe that considerations of rationality are irrelevant to religion since it operates on simple, humble, and unconditional faith in God.[96] On the other hand, many others argue that rationality is an integral part of religion, but that rationality must be understood in the proper perspective. For instance, in religion certain things might be incomprehensible, but they should not be identified with the irrational. In this context, St. Thomas Aquinas has argued that branding religion as irrational because certain religious claims cannot be substantiated by ordinary human reason is illegitimate. According to him, if we cannot understand such claims, it is not because they are irrational but because they are suprarational, just as a person's inability to see things in dazzling light arises not from a lack of light but from an overabundance of it.[97]

Kepler's overall position on this issue was very sympathetic to St. Thomas's. But Kepler went a step further to argue that religion should be rational, that it would be a contradiction for it to be otherwise. God is supremely rational, and the human being is also rational, being created in the image and likeness of God. Hence religion, which is the expression of the deep relationship between God and humankind, cannot but be rational.

Kepler's emphasis on rationality in religion should not mislead us to believe that he placed no value in religious mysteries, in feelings, and in other spontaneous expressions of the heart. He believed in mysteries like the Holy Trinity and the Incarnation. He admitted that not all truths were accessible or comprehensible to us humans and he did not consider this limitation a belittling of humans. Hence, concerning the specific nature of the planets and the different zones in which they were placed, he could advise Herwart: "Let us cease, therefore, to investigate celestial and incorporeal things, more than God has revealed to us."[98]

Critical Spirit as a Characteristic Mark of Rationality

A critical spirit was the hallmark of Kepler's rationality, especially on matters pertaining to certain Christian doctrines and practices. He refused to accept anything blindly. This critical spirit remained with him right from his school days. Already at the age of twelve he began disagreeing with the preachers who were embroiled in confessional controversies, who turned the pulpits into a launch-pad for their vendetta of malicious jealousy, rivalry, and vengeance.[99] He disagreed with their distorting Scriptures. He recounted that after these sermons he used to go to the sources and check the veracity of the interpretations he had heard from the pulpit, only to be convinced that they were unjustified distortions. Indeed, he took the teaching of the Bible (provided that it was interpreted rationally) as unquestionably true. But he did so because he was convinced that the rational God's self-revealing in the Bible would not deceive us. He was critical of all the different denominations of Christianity, which explained why he could belong fully to no single one of them. He criticized Catholicism for its upholding of the papacy and the clerical hierarchy. In Lutheranism he found the doctrine of ubiquity[100] totally unacceptable. The Calvinist doctrine of predestination he objected to vehemently.

Kepler's religion could not be fitted into the molds of any known religious denomination of his day. He accepted some doctrines and rejected some others; accepted some beliefs and rejected some others; accepted some passages of the Bible literally and some others nonliterally. Kepler's critical attitude toward the different denominations of his day and his rejection of certain key elements in each denomination were not motivated by any selfish or opportunistic considerations but by his belief that any religion should be consistent with philosophical and empirical principles. Nor did his criticism arise from a craze for novelty and innovation. In fact, he affirmed in his letter to Maestlin, written on December 22, 1616: "In general, I do not favor any teaching which cannot be found in the old Fathers of the Church. . . . Whoever is accusing me of the slightest innovation is not fair to me."[101] He did admit that religion had elements of mystery which could not be fully grasped by humans. But he opposed dogmas and beliefs which he thought gave rise to contradictions and inconsistencies. He rejected interpretations of passages in the Holy Scripture that deal with scientific matters, when they refuse to take into account developments in empirical science.

This critical spirit helped Kepler integrate science and religion. Had he blindly and unconditionally stuck to all the teachings of any of the major religious denominations as understood at that time, probably it would have been extremely difficult, if not impossible, for him to bring about such a coherent mixture of ideas. Thus his life and works show that he could coordinate science and religion into a unified and cohesive whole because he believed in an inner freedom to modify and to reinterpret some of the religious views held by his contemporaries.

For Kepler religion was not just a matter of following a set of beliefs or performing a series of rituals. It permeated his whole life and works and guided every aspect of his life-activities. As I have presented it here, Kepler's idea of religion consisted of items pertaining to God, humanity, and nature, and was governed by rationality.

At the same time, it was not immune from outside influences. These influences, especially from philosophical and scientific ideas, gave rise to his unique type of religion, a religion that could not be fitted into any one denomination.

2

KEPLER'S PHILOSOPHICAL IDEAS

Although Kepler is seldom considered a philosopher by profession, he was very much one by choice and taste. As Caspar puts it, "Kepler's *cupiditas speculandi* strove for higher things, it flew through the breadth of the world and fathomed the depths up to the boundaries which are set for mortals."[1] His inquiring mind always enjoyed asking "Why?" of everything around him. His works show that he always came up with a response to his questions; even if some of his answers turned out to be utterly wrong, he never stopped speculating. Immersing himself in philosophical issues was a delight, as is evident from his words in the AN, "When I was old enough to taste the sweetness of philosophy,[2] I embraced it all with an extreme passion. . . ."[3] Another time, when he was accused of delaying the publication of the *Rudolphine Tables*, he replied: "I beg you, my friends, do not condemn me completely to the drudgery of arithmetical calculations, but allow me to have some time for philosophical speculations, which are my only delight."[4]

If a philosopher is a person who loves wisdom rather than utilitarian knowledge or "applied science," then Kepler certainly was a philosopher. Unlike most of his contemporaries and, for that matter, many of our own, he believed that learning about nature and its laws was knowledge worth pursuing for its own sake, irrespective of its practical, immediate applications. He addressed the practical-minded with these words:

> What insensibility, what stupidity, to deny the spirit an honest pleasure but permit it to the eyes! He who fights against this joy fights against nature. . . . Do we ask what profit the little bird hopes for in singing? We know that singing in itself is a joy for him because he was created for singing. We must not ask therefore why the human spirit takes such trouble to find out the secret of the skies. Our creator has given us a spirit in addition to the senses, for another reason than merely to provide a living for ourselves. . . .[5]

According to him, knowledge is the food that sustains and enriches the soul. Indeed, the ability to speculate and to know is what distinguishes humans from other earthly creatures. His whole life was an unending quest to know more and more about the secrets of nature. This chapter discusses some of the ideas and principles that nourished and guided his ever-inquiring mind.

THE IDEA OF "PHILOSOPHICAL" AND THE IDEA OF "A REASON"

In this book the term "philosophical" refers to a collection or system of the most general and rarely revocable principles governing the structure, the operations, and our understanding of nature, and in terms of which things in nature interact and are interrelated.[6] They helped Kepler constitute his worldview.

But apart from any questions about accuracy as a representation of traditional usage (questions with which I am not ultimately concerned here), this characterization of the philosophical has a very definite use in this book. As we have seen, many commentators on Kepler have distinguished (however vaguely) the religious or theological, the philosophical, and the empirical or scientific aspects of Kepler's work. Most scholars have defined these three areas of his work with a view to asserting that one or more of them was either irrelevant or dominant in shaping his scientific contributions. A few, like Holton, have distinguished them but claimed (though only in general terms) that all three played a role. As I stated in the introduction, my purpose is to show in detail *what* those three factors involved and *how* they functioned in Kepler's thought: to show that all three were actively operative and that their operation demonstrates the unity and integration of Kepler's thought, which must not be distorted by focusing on it as an application of ideas from three independent areas. (It might even be said that this discussion will show the artificiality of the threeway distinction, even though it will be through the distinction itself that we shall come to see this.)

For these purposes, the utility of my definition of "philosophical" is that it enables us to distinguish the philosophical from the religious, on the one hand, and from the scientific, on the other, in quite specific ways that make possible the discussion of Kepler's thought in terms of the tripartite division that has played so important a role in Kepler scholarship. The definition distinguishes the philosophical from the empirical or scientific in two fundamental respects. First, philosophical principles are "most general" in the sense that they apply to nature as a whole and not, as is usually the case with empirical or scientific principles, to a limited set of cases or to a specific part of nature. Second, philosophical principles can be compromised, revoked, or otherwise violated only in exceptional circumstances (for instance, when two principles come into conflict); in particular, they are not necessarily to be abandoned in the face of any putative empirical or scientific counterevidence, and in this respect too are distinguishable from the specific laws or theories of nature which are the concern of empirical science. The philosophical may be distinguished from the religious in that philosophical principles do not necessarily involve God or anything having to do explicitly with the religious. These distinctions do not, of course, imply that the three types of inquiry are necessarily and totally independent of one another: on the contrary, the philosophical principles follow, in many cases, from Kepler's specific ideas of God, humankind, and nature as discussed in chapter 1; and empirical principles are, in many cases, shaped by philosophical principles. We shall find many other interrelations in the course of our investigations.

What I mean by "general principles" is best conveyed by stating them, and in this chapter I will discuss twelve such principles that constitute Kepler's philosophical view, or at least its salient aspects: realism, mathematizability, precision, causality, concomitant variation, dynamism, order, unity, harmony, universalizability, economy, simplicity. Some of these Kepler expressed explicitly and some only indirectly; but all of them, as we shall see in part 2, played an active role in his scientific work. Indeed, I will try to expose the dual role of these principles: sometimes they function as guiding principles in his thought and work (their role in Kepler's discovery process), and sometimes as justifications for what he had already discovered (their role in Kepler's justificatory process). Further, not all of these principles need to be satisfied simultaneously in a given instance of rational understanding, though some of them always will; and in any specific instance of reasoning, a clear violation of any of them renders that particular treatment inadequate.

From the epistemological point of view, the philosophical principles are principles of rationality in the sense that rational understanding of nature requires them. A close reading of Kepler's writings reveals two central dimensions to why he counted the philosophical principles as "reasons" in employing them to construct hypotheses or in coming to conclusions. First, the general principles are reasons in the sense that they play central roles when applied to particular problems: they specify what will count as problems, provide guidance in approaching those problems, and determine what can be accepted as adequate solutions. Second, those general principles are reasons in the sense that they are *considered*; they are not haphazard or arbitrary. They are principles that either are generally accepted (on the basis of any of a number of factors, such as proven success of the ideas, e.g., Copernicanism; their compatibility with legitimate authority, e.g., religious authority; and the testimony of persons generally recognized to be reliable, e.g., Tycho), or, when they are not generally accepted, they are principles that Kepler himself has arrived at through critical reflection. This second aspect of his counting the general principles as reasons is intimately related to his belief that they themselves are subject to critical scrutiny.

This second aspect of Kepler's treatment of the general principles as "reasons" brings out the fact that, for him, the concept of "reasoning," or of "a reason," encompasses more than just philosophical principles. Empirical and religious considerations can also count as "reasons" and, along with certain other sorts of considerations, especially logical ones, can serve as bases for critical reflection on the philosophical principles themselves. This is, of course, essential to the contention of this book: all three types of considerations entered into Kepler's thought as "reasons."

I must point out that it is not always easy to distinguish clearly one principle from some others, either in Kepler's own writings and thought, or more generally. Some characteristics overlap. For instance, the principles of unity and harmony have certain points in common. Nevertheless, adequate textual evidence shows that Kepler accepted both principles as distinct and used them as such in his scientific work. We shall therefore discuss them as separate principles.

KEPLER AND NEOPLATONISM

Kepler is often justifiably regarded a Neoplatonist because several features of this system of philosophy mark his works. This is to be expected because he was in many ways very much a product of his times. During the Renaissance, through the works of Nicholas of Cusa (1401–1464), Marsilius Ficinus (1433–1499), Girolamo Cardano (1501–1576), Thommaso Campanella (1568–1639), and others, the Neoplatonism of Plotinus, Proclus, and others had made a strong comeback. Kepler left no doubt that he was both impressed and influenced by this philosophy. In his *Harmonices Mundi*, for example, he spoke of Plato and Proclus, with deep admiration and respect.[7]

The theme of emanation is central to Neoplatonism. According to this idea, the One or the Good is the starting point. From it emanates in an eternal procession the nous (intellect), from which, in turn proceeds the soul. The material universe is the result of the procession from the soul. Concerning this process Plotinus writes:

> What are we to think of as surrounding the One in Its repose? It must be a radiation from It while It remains unchanged, just like the bright light which surrounds the sun, which remains unchanged though the light springs from it continually. Everything that exists, as long as it remains in being, necessarily produces from its own substance, in dependence on its present power, a surrounding reality directed towards the external world, a kind of image of the archetype from which it was produced.[8]

Emanation leaves the emanating being undiminished so that it remains outside of its products and yet be present to them.[9] This process is totally involuntary: whatever is full must overflow, whatever is mature must beget. As Proclus puts it: "Whatever is complete proceeds to generate those things which it is capable of producing."[10]

The process of emanation involves, not only procession, but also reversion. The products continuously go back to their source: "All that proceeds from any principle reverts in respect of its being upon that from which it proceeds."[11]

Neoplatonism maintains a positive attitude towards nature. According to Plotinus, "The visible universe is not, as the Gnostics think, evil, an unfortunate mistake, the product of some sinful affective or arbitrary whim of a spiritual being; it is the perfect image of the Intelligible World of Nous. . . ."[12] The philosophical system also looks upon the universe as an organic and unified whole.

Even though Kepler's works show that he was remarkably influenced by these and similar ideas of Neoplatonism, there was no wholesale embracing of the total system. His acceptance was limited to certain basic elements and definite principles, some of which helped him formulate his philosophical principles.

KEPLER'S PHILOSOPHICAL PRINCIPLES

Realism

The principle of realism was one of the fundamental principles to which Kepler subscribed. It holds that the world of our experience is real, not the figment of our imagination. Nor is the world any less real than the Platonic world of Ideas. Therefore to engage in the study of the structure and operations of this universe and to look for real causes to account for natural phenomena is meaningful. Realism wants science to be the study of real phenomena and scientific explanation be done in terms of real forces and material bodies.

From an epistemological point of view, this principle means that understanding gained through rational reflection corresponds to the basic structure and operations of nature. Since truth is the correspondence between our idea and the object or reality under consideration, the principle implies that truth or true knowledge is attainable. Thus realism is opposed to skepticism.

Perhaps the most forceful statement by Kepler on this point came during his heated exchange with Robert Fludd, who accused our astronomer of being a vulgar mathematician who deluded himself with quantitative shadows. According to Fludd, Kepler was concentrating on the externals and the quantifiable, whereas Fludd contemplated the internal and the essences. As he put it: Kepler "has hold of the tail, I grasp the head; I perceive the first cause, he its effect."[13] To this Kepler replied: "I reflect on the visible movements determinable by the senses themselves, *you may consider the inner impulses* and endeavor to distinguish them according to grades. *I hold the tail*, but I hold it in my hand; *you may grasp the head mentally*, though only, I fear, in your dreams."[14] Although Kepler's reply is charged with humor and sarcasm, the basic message is clear: the "tail" in one's hand about the existence of which one has full guarantee is far superior to "the head" which exists only in one's imagination or dreams and has no reliable guarantee in observation.

Having a significant impact on his scientific, philosophical, and religious ideas, the influence of the principle of realism can be seen in Kepler's statement that his whole scientific search was for discovering real causes. He wrote to Maestlin on December 14, 1604: "All my labor is to this end that I may obtain both the correct equations of the eccentric and the distances from real causes."[15] He rejected the epicycles because they were unreal. He rejected the practice of considering the geometrical center of planetary orbits as the origin and point of reference in the universe; instead, he insisted that the sun be the point of reference. One of the main reasons for this move was that he believed that the point of origin and reference must be a real body. As we shall see in part 2, Kepler made use of this principle on many other occasions in his work on Mars.

Insistence on this principle rendered Kepler's astrology reasonable and scientifically respectable. Unlike his contemporary astrologers, he refused to believe that the planets in themselves could harm or help humans. Planets are morally neutral since they are neither good nor bad.[16] According to him, the efficacy of astrology comes from the privileged status of geometrical forms (which are part of the very essence of God) and from the fact that the planets can reflect rays of light. The only point he accepted from the astrology of his day was the theory of the *aspectus*.[17] Since such formations have geometrical significance, they can be astrologically efficacious. Again, for astrological efficacy, the rays of light from heavenly bodies have to move in such a way that definite geometrical, harmonious forms are traced. Both specific geometrical forms and physical rays to trace such forms are needed. Kepler emphasized both the geometrical and the physical. As Pauli points out, the astrological effectiveness of directions that are geometrically defined in relation to the sphere of the fixed stars but do not coincide with light rays (as, for example, the direction from the earth to the vernal point) is expressly rejected by Kepler.[18] The physical requirement outlaws all unreal and magical influence of the superstitious spirits from his astrology. This consideration has prompted Pauli to postulate that Kepler made astrology "a part of his physics, indeed of optics."[19] It follows, therefore, that he did not encourage superstition but attempted to liberate astrology from superstition and give it a scientific basis, as far as the science of his day permitted. This Kepler could do because of his firm commitment to realism.

The principle of realism was mainly responsible for making Kepler a successful scientist, despite the fact that he considered Plato his "true master,"[20] and despite his strong adherence to the Platonic theory of knowledge and innate Ideas. Although subscribing to such strong Platonic views, Kepler also affirmed the reality of the world of our experience and the crucial importance of sense observations in the acquisition of knowledge of the external world. In him one sees an interaction between the realist physicist and the Platonic geometer, an interplay between realism and idealism. No traditional kind of Platonism, Kepler's brand is a strange, "unnatural" mixture of realism and idealism: realist Platonism.

Kepler differed from Plato in another way. According to the Platonists the goal of astronomy was to "save the phenomena,"[21] and this tradition gave rise to an instrumentalist conception of astronomy, as exhibited most conspicuously by the work of Ptolemy. Kepler's realism refused to go along with this Platonic and Ptolemaic tradition. He argued that the goal of astronomy was, not just to "save the phenomena," but also to explain the phenomena. Again we see the significant influence of the principle of realism on his understanding of science.

The principle influenced his religious views as well, as is evident in his debate with Hafenreffer on the eucharist. Hafenreffer said: "Where there is the word, there is his [Christ's] flesh." To this Kepler retorted: "Where there is the flesh, there is his word."[22] For his opponent, an abstract *logos* (word) was the principal element of explanation, whereas Kepler insisted on the tangible and the observable. His rejection of the Lutheran doctrine of ubiquity can be considered as a striking example of the action of realism on his religion. His sense of realism could not be reconciled with the claim that Christ is bodily present everywhere.

Why did Kepler subscribe to this principle? Or, how did he come to accept it so firmly and wholeheartedly? Copernicus must have exerted a powerful influence on him. Despite using unreal epicycles, Copernicus was a staunch realist. He believed that the sun and the planets are real bodies and that the earth along with the other planets really move around the stationary sun. He could give a convincing explanation on the basis of the real motion of the earth. As Edward Grant puts it, "The remarkable break with the medieval tradition is not to be found in the arguments Copernicus gave in support of a diurnally rotating earth. Indeed, many of these were commonplace in scholastic discussion. It is found, rather, in the insistence by Copernicus that the earth really and truly has a physical motion and in the methodological rationale which emerged from this profound belief."[23] Tycho also must have helped Kepler confirm this belief in realism when the Dane affirmed that progress in astronomy should come through a posteriori investigations rather than through mere a priori ones.[24] Furthermore, Kepler's religious views must have contributed to the origin and development of this principle: he considered God most real; the universe reflects God, and so it also must be real.

Mathematizability (Quantifiability)[25]

According to the principle of mathematizability, another major principle that Kepler vehemently adhered to, our universe is characterized by mathematizability or quantifiability, more specifically geometrizability. God created the universe by using geometrical archetypes, and hence every part of the universe has been created in accordance with geometrical principles and rules. In chapter 1, we discussed the theological or religious explanation of why nature is geometrical. Here we are concerned with the metaphysical and epistemological aspects of this mathematical character of the universe. Again and again Kepler expressed this point. "Geometry . . . has supplied God with the models for the creation of the world."[26] The traces of geometry are expressed in the world so that geometry is, so to speak, a kind of "archetype of the world."[27] The universe is mathematical.

Epistemologically, this principle asserts that what is quantifiable is intelligible, what is mathematical is rational: mathematics is the key to understanding nature. Kepler wrote that "human understanding seems to be such from the law of creation, that nothing can be known completely except quantities or by quantities."[28] In this way mathematical entities are completely knowable in themselves and other entities become known through and because of them. The same idea he expressed equally emphatically in his letter to Maestlin:

> Since God created everything in the whole world in accordance with the norms of quantity, God has given to man a mind which can comprehend quantity. For as the eye is created for colors, the ear for sounds, so also is the mind created for no other reason than to understand quantities. It will perceive an item more accurately, the closer it is to pure quantities, as though to its origin. Conversely, it will remain the more in darkness and error, the more it distances itself from quantities.[29]

The more mathematical a phenomenon, the more intelligible it is, the more open to human inquiry. Curtis Wilson proposes a similar view when he writes of Kepler as presenting a "new way of seeing the world as a connected system. This invisible connection between bodies, the radiations of light and heat, and motive force, with Kepler became quantifiable, and in this respect alone, he believes, graspable by man."[30] Hence for our astronomer theoretical invisible entities like radiation and forces become intelligible because they are quantifiable. The extreme importance he attached to this principle can be seen from the reply he gave to Fludd when the latter accused Kepler of using mathematics in the study of nature. "Without" mathematical demonstrations, Kepler said, "I am like a blind man."[31] Mathematics gave him the light to see, the illumination to understand, the key to unlock the secrets of nature.

That there is a close relationship between mathematics and the universe was nothing new. The Pythagoreans and the Platonists, for instance, subscribed to this view. However, controversy existed about the nature of this relationship and about which branch or aspect of mathematics was involved. The Pythagoreans long ago had argued that the universe was essentially mathematical, that everything in the universe could be accounted for in terms of mathematics; the relationship between the universe and mathematics was an essential one, and, according to them, arithmetic was the fundamental form of mathematics and hence theirs was an arithmetical or number theory. Plato also believed that the universe was mathematical in nature, but he emphasized geometry; Plato's emphasis on mathematics was so well known that the sixteenth- and seventeenth-century revival of mathematics in physical science was often referred to as the resurrection of Platonism, what Koyré calls *la revenche de Platon*. Aristotle did not reject mathematics altogether—it did have a crucial role in theoretical astronomy—but he banned it from physics, from the scientific study of nature. For him and for his followers science was knowledge in terms of the essences; mathematics referred to quantity, which was an accident, so it could not yield true scientific knowledge, and mathematics had no real role in the physical or scientific study of nature. In recent times Stillman Drake has tried (unsuccessfully) to reinterpret this position and to cast Aristotle in a brighter light.[32] Whatever be the merit of Drake's attempt, the fact remains that at the time of Kepler, the Aristotelians had banned mathematics from physical science. By emphasizing mathematics' role in the study of nature, Kepler was going against a strong and prevalent tradition of his day.

Religious people also had condemned the use of mathematics in science. Luther, for instance, declared it as the "enemy of all theology."[33] Applying quantity to God and his creation would have been anathema for Luther.

However, others differed from the prevalent traditional Aristotelian view. The Merton school, Buridan, Oresme, Cusa, and Copernicus were a few among those who advocated a legitimate and beneficial role for mathematics in the study of nature.

I believe that Kepler in some ways agreed with his predecessors but went far beyond them all. He remained very much a transitional figure in this respect as well. He contributed decisively to the development of modern mathematical science. In fact, he was the first real mathematical physicist, in the modern sense of the term.

Right from the beginning of his career, the mathematical worldview captured his rational and critical mind. He was so thoroughly intoxicated with this idea that in a way it became a be all and end all for him. Perhaps no other person in recent times, with the probable exception of Descartes, has given as prominent a place to mathematics as Kepler did. For Galileo it was the language in which the book of nature was written, i.e., it was only the way in which we describe nature. But for our astronomer, nature itself was formed according to mathematical laws.

Although Kepler emphasized mathematics more strongly than all his contemporaries, labeling him a mathematical mystic, as many scholars like Burtt have done, is inaccurate and unjustified. Kepler's attachment to mathematics showed certain significant differences from mysticism. Mystical experience is often referred to as "ineffable" since it cannot be verbalized or communicated adequately to others; one reaches the mystic heights by distancing oneself from rationality more and more, and mysticism usually implies an excessive role of emotions and irrationality. Some of Kepler's views were excessive and turned out to be incorrect, but he cannot be accused of emotionalism and irrationality. He had good reasons for emphasizing mathematics. He arrived at his ideas about mathematics by well-reasoned steps and he was not afraid to accept limitations to mathematics when reason demanded.[34] He justified his position not only on religious but also on empirical considerations. Again, unlike the mystics, he could communicate his ideas to others who understood what he said, although they did not always agree with him.

In his mathematical view of the universe Kepler subscribed to the Platonic rather than to the Pythagorean tradition. For him, nature is geometrical. Pure numbers or pure arithmetic have no special place in his system. Perhaps the collapse of the Pythagorean number theory of the universe in the face of incommensurability problems (e.g., the ratio of the diagonal to the side of a square is not expressible as a "rational" number) influenced him to take his position, as it undoubtedly did Plato. This should not surprise us because, as we have discussed earlier, the universe for Kepler is a unified whole. If arithmetic is a true characteristic of the universe, it must be applicable to all cases. The incommensurability problem betrayed the nonuniversal applicability of arithmetic.

Kepler had other arguments for geometry and against arithmetic and number theory as well. For instance, the properties of number were accidental (except the number three of the Holy Trinity), whereas those of geometry were grounded in nature.[35] By calling the number theory accidental he meant that it could not bind its conclusions with any intrinsic necessity. For example, according to the number theory, the number of planets can be six or twenty-eight; there is no intrinsic reason to choose either one in preference to the other. On the other hand, geometrically speaking, the number of planets must be six, since there are only five regular solids.

In his emphasis on geometry, he differed significantly from several of his contemporaries. For example, George Joachim Rheticus wrote about six, the number of the planets: "Who could have chosen a more suitable and more appropriate number than six? By what number could anyone more easily have persuaded mankind that the whole universe was divided into spheres by God the author and creator of the world? For the number six is honored beyond all others in the sacred prophesies of God and by the Pythagoreans and the other philosophers."[36] Brahe had strongly advised Kepler to explain astronomical findings in terms of the number theory: "For there is no doubt that everything in the universe has been related and ordained by God in accordance with fixed harmony and proportion, in such a manner that it may be represented just as well by numbers and figures, as was foreseen formerly to some extent by the Pythagoreans and the Platonists. Therefore, direct the power of your mind to the number, and if you find perfect agreement without the least defect or deficiency, then you will be in my view a great Apollo."[37] Kepler rejected the suggestion. In his view, only geometry existed before the creation of the world. In fact, geometry was coeternal and coexistent with God. Arithmetic or number could claim no such unique prerogative: number came after the world was created. "God created the bodies of the world in a determinate number, but number is an accident of quantity (*quantitatis accidens*), number, I assert, in this world. For before [the creation of] the world there were no numbers except the Trinity who, of course, is God himself."[38]

Kepler's arguments show that his rejection of arithmetic and his wholehearted acceptance of geometry were based on his idea of God and God's essence. His God is a geometer, not an arithmetician. His religious views could lead him to significant philosophical ideas.

Kepler had a rather restricted concept of the geometrical. Namely, what was geometrical was whatever was constructible by ruler and compass, because only such figures were constructible by natural means and God used only natural figures in the construction of the universe.

Can we find other reasons for his strong adherence to the supreme importance of geometry? Many scholars point out Cusan influence. I do not deny that this view has justification, for Cusa did say: "We have no certain knowledge except mathematical knowledge and this latter is a symbolism for searching into the works of God. Thus, if great men said anything important, they based it upon a mathematical likeness. . . ."[39] However, for the Cardinal, mathematical entities were logical entities produced by our mind. As he explained, "The mathematical [entities] are neither an essence nor a quality, rather they are notional entities elicitated from our reason."[40] This position is a far cry from Kepler's idea that mathematics is the essence not only of the human but also of the divine mind. Hence Cusa's influence on Kepler could not have been decisive.

Perhaps Proclus, the Neoplatonist, had a deeper influence, as is evident in the HM, where Kepler quoted Proclus at great length.[41] Proclus wrote: "Therefore we must submit that the soul is the source of mathematical categories [specierum] and ideas."[42] Kepler's remarks on this passage left no doubt that he made Proclus's view his own. In fact, he wrote in the margin: "The true essence of mathematical entities is in the soul."[43] Copernicus's theory must have convinced Kepler of the supreme power of mathematics. Copernicus showed that by using mathematical reasoning he could account for real phenomena: the number, ordering, and size of the planets could be derived as a matter of course. Despite these clear influences, I think that an original, creative mind like Kepler's could not have been so overpowered by one or a few persons, however eminent. They may certainly have had some influence, but they could have been only partially responsible.

I think that the main reason for Kepler's emphasis on mathematics consisted in this, that he had a deep insight into its potential. He realized that it could provide him with the best means to achieve his goal of a unified worldview. It provided the cosmic connection he was looking for. It provided the link between God, humankind, and the universe: geometry is part of the essence of the divine mind; humanity is the image of God because the human mind shares in the geometrical essence of God and is able to understand geometry; the universe is the image of God because it is created according to the laws of geometry. In short, geometry gave him a system that had its ultimate source in God. More than anything else, geometry could help him set up the kind of system he wanted to build, and, in turn, mathematics could provide the life-blood for his system.

Kepler had some empirical proof as well. In his letter to Herwart at the end of January, 1607,[44] Kepler explained how he came to give such a unique place to geometry. In that letter he proposed two axioms which basically argued that corresponding to every geometrical consonance, there was an acoustical (musical) consonance; corresponding to every geometrical, abstract relation there was an empirical, concrete relation. He developed his position as follows: Assume that a string has been divided into several sections of a circle. It is observed of stretched vibrating strings that the smaller the section of string, the faster it moves. Again, faster motion produces a more shrill sound, whereas a longer portion with slower motion gives a deep sound. This shows a relationship between the ratio of the length of the string considered and the kind of audible sound produced. In other words, the mathematical ratios and the musical sounds produced are interrelated. Similarly, certain definite sections of the circle are in consonance with the circle, i.e., certain ratios of a section of the circle to the whole circle can give rise to a consonance. The Pythagoreans also gave a similar empirical basis for their arithmetical theory. But according to Kepler, this happens *only* when the ratios involved are geometrical; when the ratios are nongeometrical, dissonance is the result. These findings lead to the conclusion that whatever harmonizes in geometry in an abstract way, harmonizes in nature in a concrete way. Hence there is a one-to-one correspondence between the geometrical and the natural. In Kepler's view this comes about because geometry is the archetype of the universe.

My claim that he was very much a transition figure who departed from the old ideas in significant ways, even while he subscribed to some of them, is conspicuous in this context. In some ways he agreed with both Plato and Aristotle, but in some other matters he differed from them. With Plato he believed that mathematics was the key to unlocking the secrets of the universe. But he disagreed with Plato by subscribing to the idea that our material universe is perfectly real and we can have firm and reliable knowledge of it. Unlike Plato, Aristotle believed that the material world is real, but he discouraged, if not banned, the use of mathematics in the study of physical nature. Kepler combined Plato's idea of mathematics and Aristotle's realism and thus contributed a true mathematical physics. In this way we see Kepler having both the old Platonic idea of mathematics and the modern idea of mathematical physics. In arriving at mathematical physics, his religious idea of a geometer God and his philosophical idea of a geometrical universe were crucial.

CONTROVERSY WITH FLUDD

Some of Kepler's ideas about geometry and quantity come into very sharp focus in his controversy with Fludd. Fludd was a staunch Aristotelian who used all the weapons in the age-old Aristotelian arsenal to shoot down Kepler's adherence to the principle of mathematizability. Fludd argued that mathematics could never put us in touch with the real essence of things but could only delude us with quantitative shadows.

According to Fludd, Kepler's was the spurious and blundering type of mathematics in which nature remained hidden. On the other hand, his was the formal mathematics in which nature was measured and revealed. This formal mathematics could yield essential, rather than superficial, knowledge, "so that the mysteries of science having been revealed, that which is hidden may become manifest and that the inner nature of the thing, after the outer vestments have been stripped off, may be enclosed, as a precious gem set in a gold ring, in a figure best suited to its nature—a figure, that is, in which its essence can be beheld by eye and mind. . . ."[45]

Fludd's Aristotelianism is quite evident from his criticism that Kepler's mathematics could not yield essential knowledge, since mathematical knowledge was nothing but knowledge in terms of the accident of quantity. Kepler, on the other hand, considered quantifiability, mathematizability, a criterion of knowability. Indeed, as Pauli points out, for Kepler "only that which is capable of quantitative mathematical proof belongs to objective science, the rest is personal."[46]

With regard to the formal mathematics of his opponent, Kepler declared: "You, Robert, may have for yourself the glory of it and that of the proofs found therein. . . ."[47] He added that his concern was the visible movements amenable to empirical study. Clearly Kepler has already taken the decisive step away from Aristotelianism to modern science. No more the concern with the invisible and incomprehensible essences, rather there emerges the visible movements determinable by the senses. But has he become a positivist? Has he given up the search for the essential or the real? By no means. For Kepler being geometrical is an essential element of being real, so knowledge revealed by geometry cannot be regarded as accidental or superficial. This debate brought to the fore the sharp contrast between Kepler's and traditional views, thereby revealing his remarkable originality.

Precision

The principle of precision is closely related to that of mathematics insofar as mathematics is the paradigm of precision and accuracy. From a metaphysical point of view, this principle asserts that nature is precise, nature follows precise laws. Every detail of nature is ordained and arranged exactly and precisely; nature has not been created haphazardly or approximately. An important epistemological aspect of this principle is Kepler's view that the indefinite and the infinite are unknowable. This is a natural consequence of this principle because the indefinite and the infinite do not lend themselves to precise and accurate determination. Thus with the Greeks he shared the "horror of the infinite" and of the indefinite.

Kepler used this principle of precision repeatedly in his work, especially in his search for accurate results. Although he seemed to be convinced of the truth of the polyhedral theory, the inaccuracy it betrayed when the theoretical values were compared to the observed ones disturbed him persistently. He unhesitatingly confessed that it was with the intention of getting a more accurate confirmation of the theory that he went to Tycho. Whenever he proposed a new theory, he insisted on accurate agreement with observed data.

This principle had theologically significant consequences also. The belief that human knowledge is capable of attaining mathematical accuracy rendered human knowledge and divine knowledge qualitatively indistinguishable since both would be equally precise.

And so comes to the fore one of the disagreements between Plato and Kepler. Plato argued that a precise knowledge of the changing world was impossible because of the inherent transitory (and hence less real) character of the universe and because of the inability of the Demiurge to create a geometrically precise universe. The principle of precision clearly opposed such a view. By the principle of realism, for Kepler the world of becoming is as real as anything can be, and by this principle it follows that we can have a precise knowledge of that world. Again, his idea of the *optimus creator* ruled out any imperfect or inaccurate creation. He therefore opposed Plato both from realist and religious points of view.

How did Kepler come to believe this principle? It can be looked upon as a direct outcome of his geometrical worldview. Given his geometrical understanding of God, humanity, and nature, the principle of precision naturally followed. Later on he had to extend the basis of the principle beyond the rules of geometry to include harmony as well.

Causality and Sufficient Reason

The principle of causality states that every event has a cause. Nothing in the universe happens without a cause. One may or may not be able to identify the exact cause of an occurrence, but it must have one. In science, a satisfactory explanation of a phenomenon must be in terms of its real causes.

The epistemological side of this principle is the principle of sufficient reason, which stipulates that there must be a sufficient reason for every event in nature. Nothing happens by chance. Therefore the principle bans chance events from nature. In fact, according to this principle, rational understanding consists in identifying the sufficient reason for every happening. Since both the metaphysical and the epistemlogical aspects of this same basic principle are extremely important, we shall discuss both in some detail.

Kepler emphasized the principle of causality very much in his scientific work. This fact is borne out strikingly by the title he gave to his most scientific work: "New Astronomy or Physics of the Heavens Explained on the Basis of the Laws of Causality and Developed in Analysis of the Movement of Mars Based on Observations of Tycho Brahe." The law of causality was central to his investigations of the motion of Mars. As he wrote to Maestlin, all his labor aimed at discovering the true causes of phenomena.

Emphasis on the discovery of causes was at least partially responsible for the creation of the new astronomy or new science. The emphasis transformed astronomy or science as an enterprise "to save the phenomena" to one "to explain the phenomena" since for Kepler identifying the cause and causal connection was an integral part of a scientific explanation.

In discussing causality let us be clear about the different kinds of causes and the particular type of cause a scientist or thinker emphasizes. Most scholars agree that the shift from medieval to modern science was characterized by a shift from stressing final cause (teleological cause) to emphasizing efficient cause. I believe that here, too, Kepler was very much a transitional figure, that one can see in his works emphasis on both the final causes and the efficient causes. Although he never gave up the importance of teleological explanation in his science, he did search for explanation in terms of efficient causes. Obviously, he contributed in a significant way to science's transition from the medieval to the modern, and he deserves to be counted among the founders of modern science.

Many Kepler scholars fail to recognize the importance he gave to efficient cause. Rudolf Haase, for instance, believes that Kepler thought teleologically, emphasizing final causes.[48] Burtt attributes some novelty to Kepler's thought because the astronomer "thinks of the underlying mathematical harmony discoverable in the observed facts as the cause of the latter, the reason, as he usually puts it, why they are as they are."[49] The observed facts are produced by the inner, deeper harmony. One may say that the inner, deeper harmony is the efficient, rather than the formal or final, cause of the observed facts. However, Burtt thinks that harmony was not a kind of efficient causality but was substantially the "Aristotelian final cause reinterpreted in terms of exact mathematics."[50] And so Burtt takes back the compliment about the novelty of Kepler's view of causality with the other hand.

Pauli certainly makes an effort to show how efficient causality played a significant role in Kepler's thought and work. For instance, as we have already seen, he interprets Kepler as holding the idea that even for astrology to be effective light rays have to operate as efficient cause. Pauli, however, does not elaborate on this point.

Perhaps the most conspicuous instance of Kepler's shift from explanations in terms of formal or final cause to explanations in terms of efficient cause was his move from *anima motrix* to *vis motrix*. At first in his MC he explored planetary motion as being caused by an *anima motrix*. In the Aristotelian system such a cause usually refers to the formal, or in some cases, to the final, cause, because a body's ability to move can be considered part of the very essence of the *anima*. On the other hand, *vis motrix* is identified with an efficient cause. The *vis motrix* is not what the sun is, but is that by which the sun brings about planetary motion. This move highlighted not only the importance of efficient cause but also marked the beginning of explanation in terms of mechanical causes, thereby ushering in the era of the mechanical philosophy.

Kepler's emphasis on efficient and mechanical causes can also be seen in his elucidation of the noncircular motion of Mars in AN. Here he compared the planetary motion to the motion of a boat in a circular river, the deviation from circularity being brought about by the orientation of the boat with respect to the current and by the twisting of the oar in the appropriate direction. His explanation of planetary motion by using magnetic forces can also be understood along this same line.[51] In the light of these considerations Kepler can hardly be considered a late medieval figure still imprisoned in the traditional stress on formal and final causes. He was still away from the fully modern scientific view of causality; nevertheless he had already made his move along that path.

The principle of sufficient reason Kepler accepted wholeheartedly. Chance had no place in his universe. Since no reasons have been or can be assigned to a chance event, such an event would be unworthy of a rational God, would be an affront to the all-rational, all-knowing, and all-powerful creator. On several occasions Kepler proclaimed: "The creator chose nothing by chance."[52]

The principle of sufficient reason is often associated with Leibniz because he developed it and successfully applied it to science and philosophy. However, already more than three-quarters of a century before him, Kepler had proposed and consistently made use of this principle. Again and again he used it in his study of the motion of Mars. Until and unless he found a sufficient and satisfactory reason for the question under investigation, he refused to proceed to the next question. For instance, in AN as soon as he had discovered that the orbit of the planet was not a circle but an oval, he wanted to find out the reason why nature chose an oval path. Often such reasons he identified with causes; in this case he looked for the causes giving rise to the oval path. Again, in HM, the search for sufficient reason led him to the discovery of musical harmony. He could not see why God should have chosen the numbers 1, 2, 3, 4, 5, 6 for generating musical consonances while excluding numbers 7, 11, 13, etc. He looked for sufficient reason for this puzzling observation, and his search initiated and spurred by this principle turned up an explanation in terms of the "just" tuning that had just been discovered in his own lifetime. The seven ratios of the just scale are: 1/2, 2/3, 3/4, 4/5, 5/6, 3/5, 5/8. He found that the entire just scale or system of harmonic relationships could be derived from four regular polygons and their derivatives. This showed that the numbers 1, 2, 3, 4, 5, 6 were related to harmonic ratios, whereas 7, 11, 13, etc., were not. Thus he believed himself to have traced the sufficient reason for God's choice of these particular numbers for generating musical consonances.[53] Furthermore, since nothing happens by chance, if he found agreement between theory and observation, even an imperfect one, he would argue that the agreement could not be accidental and would explore further to find the reason for the agreement. In MC, after having noticed the agreement between the polyhedral theory and the observed data, he noted: "Hence you can realize how easily it would have been noticed, and how greatly unequal the numbers would have been, if this understanding had been contrary to nature, that is, if God himself at the creation had not looked to these pro-portions. For certainly it cannot be accidental that the proportions of the solids are so close to these intervals."[54] Kepler was convinced that the remarkable agreement between his theory and Copernicus's observational data could not have been by mere chance.

This principle must have stood in good stead during his long and laborious investigations of the orbit of Mars. So often the agreement between his theory and observational values was quite rough. But he believed that even a rough agreement could not come about by accident. He looked upon a rough agreement as the first ray beckoning him to the full brightness of a true theory.

Concomitant Variation

Basically, the principle of concomitant variation states that the concurrent existence, appearance, or disappearance of certain characteristics with a phenomenon usually admits of a causal interpretation between the phenomenon and the characteristics. The Keplerian version of this principle was given in AN when he discussed the cause of planetary motion: "Now a well-known axiom in natural philosophy states that the phenomena produced at the same time and in the same way, and which are attenuated to the same degree, are [either] interrelated, [or] result from a common cause."[55] Obviously, this principle is based on the principle of causality since it argues that from the same cause under the same conditions the same results should follow. Hence, just as in the case of the principle of causality, here too the epistemological dimension is given by the principle of sufficient reason.

Dynamism

The principle of dynamism says that nature is not static, but dynamic. Nature is not passive, but active. Nature is ever changing, ever developing. This principle gives rise to a dynamic worldview, as opposed to the static outlook of the Aristotelian system, which would have everything having its "natural place" or striving towards it. The dynamic principle also emphasizes the importance of forces as bringing about changes in nature because force is an external manifestation of dynamism.

The principle of dynamism is also an aspect of causality since dynamism stipulates that everything has a cause and that that cause is active, capable of producing real effects. Just as the principle of sufficient reason is the epistemological side of the principle of causality, sufficient reason is also the epistemological part of this principle of dynamism, insofar as there is an active aspect of the principle of sufficient reason.

One of the important epistemological implications of this principle is that it affirms that we can have a science of the changing world. Plato had argued that the world of our experience was in constant flux and therefore we could not have firm, reliable knowledge of it. This contrasting principle of nature muffles the Platonic argument, since change and dynamism of the universe are as real as the world itself and so there can be a science of the constantly changing world. It kept Kepler optimistic concerning the possibility of reliable knowledge about the operations of our universe, despite his strong allegiance to many other Platonic ideas.

The influence of this principle can be seen in his search for the forces responsible for planetary motion. Belief in the dynamism of the universe directed him to look for dynamic agents responsible for planetary motion. He found the *species*[56] emanating from the sun to be such an agency.

This principle was deeply rooted in his conception of God as substantial energy. If God is substantial energy, then one has to expect the universe, the reflection of God, to be active and dynamic also. This philosophical principle had significant scientific consequences rooted in Kepler's religious ideas.

Order

The principle of order says that despite the multiplicity of beings, despite the many forces and varied activities in the universe, there is genuine order in it. Order comes about not simply because the objects in the universe are in their right place, but also because the objects and the operations are governed by true laws of nature. The principle has important epistemological dimensions. For one thing, it makes prediction of future phenomena possible, for if natural objects and operations follow definite law and order, then one can predict their future, provided the law is known. Another side, which Kepler himself explicitly mentioned, is that the chaotic and the disorderly are unintelligible; thus he argued that very little order was to be found within the sphere of the fixed stars and so we could know very little about that sphere, and, in fact, he did very little research on the fixed stars.

Since antiquity, people had believed that the universe is characterized by order. Plato and Aristotle, for instance, subscribed to this view—one of the main reasons that they both rejected atomism, a fundamental tenet of which was that atoms moved in a random and chaotic way. The atomists tried to explain the operation of the universe and the formation of various things in it in terms of atoms in constant and chaotic motion. Both Plato and Aristotle considered such a view ridiculous. To be sure, in the *Timaeus* Plato talked of an ordered universe emerging from primeval chaos. But this chaos was prior to the emergence of the universe; once the universe was formed, order prevailed all over it. No doubt, the concern for order was his prime motivation for stipulating circular motion in astronomy, because the circular path is the most regular and orderly and always comes back to the point of origin. Aristotle also stressed order and unity based on right arrangement, which consisted in everything being in its right place and so was a static order. The idea of order arising as a result of the universe obeying dynamic, physical laws was not present.[57]

Kepler, on the other hand, emphasized this latter, dynamic kind of order. For the Aristotelians, being dynamic was a potentiality, part of the substantial form of the being. Motion, a manifestation of dynamism, was directed at bringing the being to its natural position. Having rejected the doctrine of natural place, Kepler had to explain order in the universe differently. According to his new and original view, the order in the universe has its source in the fact that the universe is governed by natural laws, kinematic and dynamic laws, laws that are mathematical in character, laws that can yield a precise account for every detail in nature. His life ambition was to discover a cosmic law binding the universe together in an orderly fashion. It follows, therefore, that neither the overwhelming multiplicity in nature nor the teeming activities of nature in any way render the universe chaotic or disorderly.

His belief in cosmic order had a decisive influence on Kepler's scientific work; all through his scientific research he looked for this order. Even when things looked confused, he never gave up his struggle because he was convinced that chaos and confusion were only superficial and real order lurked behind. This principle must have played an important role too in Kepler's acceptance of the Copernican theory since it presented a dynamic and ordered universe in which the ordering and distance of the planets followed as a natural consequence of the hypothesis of the moving earth and stationary sun.

From the epistemological point of view the principle of order had tremendous significance in Kepler's work, since the principle of predictability is based on the principle of order. Predictability has always been a crucial criterion of truth for a scientific theory. For Kepler also this was a key criterion. He himself admitted in MC that one of the main reasons that he accepted the Copernican theory was its ability to predict future motions more accurately than its rivals.[58]

Kepler's acceptance of the principle of order also had a religious foundation. In God, perfect order always prevails; God never acts at random. Hence it is no wonder that the universe created by God and a reflection of God should be characterized by order.

Unity

The principle of unity was another philosophical idea that Kepler clung to tenaciously. As Caspar remarks, it seized the astonomer and influenced him in everything he did.[59] According to this principle, the universe is a unified whole; it consists of myriads of beings and is dynamic and abounds in activities, but it is nevertheless orderly because of the underlying unity. Epistemologically this principle expresses itself in the belief that an adequate explanation consists in reducing many phenomena to one or a few factors. Hence the principle and practice of reduction, which tries to account for many phenomena in terms of a single, or at most a few, explanatory facets, has its foundation in the principle of unity. If the universe is characterized by unity, then it becomes possible to carry out such a reduction.

Aristotelian astronomers and others, too, had emphasized unity in nature, but that unity was of a limited, parochial kind: beings in the terrestrial sphere exhibited a certain degree of unity among themselves. Similarly, those in the celestial sphere showed unity among themselves. But the two constituted two different worlds, following two different sets of rules. Between the terrestrial and the celestial was no unity. Furthermore, the limited nature of this sort of unity was evident, not only between the terrestrial and the celestial realms, but even within each individual realm. In the celestial sphere, in astronomy, this was very obvious. The Ptolemaic system, which had dominated the astronomical world for over fifteen hundred years, failed, despite all its mathematical elegance, to give a unified view. Copernicus lamented this situation and criticized the Ptolemaic system for producing a "monster."

Kepler, on the other hand, believed that a unifying principle exists behind the multiplicity of beings in the world. In this regard he seems to have been influenced by Cusa. As Westman points out, "What excites Kepler in the Cusan thought-system is not his cosmology, which he may well not have known, it is rather the frequently re-iterated epistemological–metaphysical claim that behind multiplicity and opposition there lies a unifying principle which is the object of all rational inquiry."[60] Surely Kepler had a sharp and analytical mind that could penetrate the depths of complicated mathematical analysis, yet he always attempted to see each element in the world as linked to everything else: an item isolated and detached from everything else failed to impress him. His unit of thought was usually not a single, detached, individuated idea but a network of interrelated ideas. This may explain the long and complex sentences he often used in his writings, where each sentence had so many parts, different parts talking about different but interrelated points, which, obviously, rendered his language and style difficult to understand and still harder to translate.

Since Copernicus also emphasized the importance of a unified universe, from this point of view he did have an influence on Kepler. However, the Canon subscribed to the separation between the celestial and the terrestrial.[61] He attributed a natural rotation to the earth and thus ascribed to it a property reserved only for the planetary bodies in the heavens. But he did not seem to have taken this theory as a counterargument against the age-old dichotomy between the celestial and the terrestrial. Rather, he explained this rotation as a natural consequence of the sphericity of the earth. Kepler's important contribution was that he argued that no such separation existed. His firm belief in cosmic unity demanded that the two realms not be split in any way. Both constitute a single unified whole, and so laws applicable to the one must be equally applicable to the other. His belief in unity enabled him to develop a celestial mechanics on the model of the terrestrial.

The idea that the universe is an ordered and unified whole also helped Kepler develop a criterion for theory choice. If the different parts and laws of the universe are interconnected, then a true theory should explain not only the specific phenomenon it was originally intended for, but also others. For instance, he argued that the Copernican theory was superior to the Ptolemaic because it not only gave the distances but also the relative ordering and number of planets. Explanatory fertility, i.e., the ability to explain items other than those originally anticipated, was a criterion for Kepler, and this was also an outcome of the principle of unity.

His belief in ordered unity undermined the notion that there existed only one, or a few, privileged sources of truth. To a Christian religious fundamentalist, the Bible is the most privileged, if not the only, source of truth. An empiricist, on the other hand, can accept primarily only the findings of the senses. Kepler was liberated from such a narrow choice regarding the source of truth. For him, truth could come from any side and he was prepared to accept it irrespective of its origin and lineage. He could write to a friend: "I ask God to make my spirit strong so that I direct my glance at the pure truth, from whichever side it should be presented. . . ."[62] If the universe is a unified whole, then it is not important whence knowledge or truth comes. Thus in his system he could accept knowledge from revelation, geometry, empirical data, religious beliefs, historical documents, etc. This philosophical position was at the basis of Kepler's synthetic view.

This principle also had its basis in his religious beliefs. God, despite the trinitarian character, is one; God is absolute unity. Unity must be present in the universe, God's reflection.

Harmony

The principle of harmony, according to which our universe is a harmonious whole, was another idea that had an overwhelming influence on Kepler. The different parts of the universe are related to each other in a harmonious way, pleasing to the heart and mind. God fashioned our universe not only according to the canons of geometry, but also of harmony. Epistemologically, the principle of harmony implies that our understanding of the universe needs not only to consider geometry but also harmony, the complementary facet to the geometrical nature of the universe.

Nothing could bring so much joy and peace to Kepler's heart as the contemplation of the harmony of the cosmos. As Caspar says, "These were the thoughts to which he clung during the trials of his life and which distributed light to him in the darkness which surrounded him. They formed the place of refuge, where he felt secure, which he recognized as his true home. . . ."[63] "The magic of the word harmony transported him to another, a pure, paradisiacal world."[64] This idea had an incomparably unique place in his thought and works. As he himself exclaimed: "I feel carried away and possessed by an unutterable rapture over the divine spectacle of the heavenly harmony."[65] His fascination for harmony was so all-conquering that he wrote, in his letter to Heydonus of London in October, 1605: "May God deliver me from astronomy, so that I can devote all my attention to my work on the harmonies of the universe."[66] Kepler was not unaware of the limitations of harmony and of the results he obtained. For instance, in HM he admitted that the heavenly music he discovered as a result of years of strenuous work was inaudible. Indeed, he accepted that no voices existed in heaven. Still more, he was fully aware that he was considering the movements solely as apparent from the sun and therefore he was dealing with motion as it would be observed by someone from the sun, not from the earth.[67] And even so he exclaimed: "I do not know why but nevertheless this wonderful congruence with human sound [*humano cantu*] has such a strong effect on me that I am compelled to pursue this part of the comparison also, even without any solid natural cause."[68] Indeed, the thoughts and ideas of harmony had made such a captive of him that he was almost helpless. It was not AN, his most scientific book, but HM that he considered his best and greatest book, his greatest achievement.

The idea of harmony is ancient. For instance, heavenly harmony, i.e., that the distances of the planets ought to be arranged in a pleasingly harmonic pattern, in simple ratios or "proportions," goes back at least to the ancient Greeks. The Pythagoreans believed in the harmony of the spheres, according to which each heavenly body emitted a sound and these sounds together made a harmonic music.

They knew that the sound's pitch depended on the speed of the vibrations of a moving body (string, air column, celestial sphere, etc.): the faster the motion, the higher the pitch.

Plato also talked about the harmony of the universe, as is evident from the *Timaeus*, where the Demiurge created the universe by harmoniously arranging the different elements.[69] Aristotle, too, spoke of cosmic harmony.[70] For our purpose the most significant ancient authority to discuss heavenly harmony in detail was Ptolemy. In fact, Kepler himself wrote about it in HM. After long attempts to get hold of a copy of the *Harmonices* of Ptolemy, Herwart sent Kepler one to read. And his reaction? He exclaimed: "There, beyond all my expectations and with the greatest wonder, I found almost the whole third book given over to the same contemplation of celestial harmony, fifteen hundred years ago."[71] Of course, he clarified that the similarity between his ideas and Ptolemy's was evident only in the questions asked, not in their solutions, because in Kepler's view Ptolemy's work "seemed to have recited a pleasant Pythagorean dream rather than to have aided philosophy."[72] Nevertheless, Kepler was pleasantly surprised at the sameness of concern between him and the great ancient astronomer and took this as an assurance that "the finger of God" was guiding him.

The music of the spheres was a very popular concept in the Middle Ages and remained so in Kepler's time. According to this concept, God had arranged the planets in a pleasingly harmonious pattern so that the heavens could sing his praises. Obviously, this belief was based on, or at least was strengthened by, the words of the Psalmist: "The heavens proclaim the glory of God."

Different views have emerged as to what harmony meant to Kepler. Westman distinguishes two kinds of harmonies: microcosmistic and architectonic. Microcosmistic harmony "refers to the principles of analogy or similarity of structures between the invisible and visible realms." On the other hand, architectonic harmony "denotes the principles of mathematical beauty and simplicity—completeness, equality, regularity, symmetry, the concordance of whole and parts—which are themselves reflected in the visible world."[73] The first kind, therefore, consists in an agreement between an original and a copy, whereas the second kind comprises a cluster of related ideas. The second kind seems to defy any definite and precise description, a difficulty that also attends Wilson's attempt to specify what harmony is by giving only a possible list of connotations of the term: "Simplicity and symmetry of geometrical arrangement, simplicity of causal explanation, and the architectonic beauty of system created by an artist God."[74] This partial list is very similar to Westman's description of architectonic harmony, which also treats of harmony in terms of simplicity, symmetrical arrangement, right fit, etc. Holton, quoting H. Zaiser, attempts to elucidate the matter as follows: "Harmony resides no longer in numbers which can be gained from arithmetic without observation. Harmony is no longer the property of the circle in higher measure than the ellipse. Harmony is present when a multitude of phenomena is regulated by the unity of a mathematical law which expresses a cosmic idea."[75] This passage seems to identify harmony with mathematical reducibility, in the sense that harmony exists where various phenomena can be expressed by a mathematical formula. Only a partial description of harmony, mathematical reducibility focuses on only one of harmony's aspects. This point becomes clear when we study what Kepler himself said about his idea of harmony.

In HM Kepler gave a detailed, though not so clear, discussion of harmony.[76] He talked of two kinds of harmony: sensible harmony (*harmonia sensilis*) and insensible harmony (*harmonia insensilis*). The two are intimately connected because the latter is but the archetype or *paradigmatio* of the former.[77] According to him, sensible harmony involves four elements: (1) two sensible items of the matching kind that can be compared on the basis of quantity; (2) the soul that can perform the comparison; (3) the inner reception of the sensible; (4) the appropriate proportion or ratio that is understood as harmony. Harmony, therefore, has objective and subjective facets. Agreeable or appropriate, quantifiable relation or proportion constitutes the objective aspect. The mind and the inner reception of the appropriate proportion by the mind make up the subjective aspect. The inner perception of the appropriate mathematical proportion gives rise to sensible harmony. Sensible harmony for Kepler is not a purely intellectual idea but appeals to our aesthetic feelings as well. According to him, our souls have the inborn ability to know what the agreeable proportion is and what is not. The soul is born with archetypal harmonic laws.

Thus an essential element of harmony is the appropriate, agreeable, pleasing, quantifiable relation or proportion between the items under consideration. What constitutes the appropriate or the agreeable is not clearly defined in Kepler. It may come from simplicity, from symmetric arrangement, from regularity, from unifiability, from reducibility, etc. The element of appropriateness is open ended. This open endedness may explain the lack of definiteness in the description of harmony by scholars like Wilson and Westman (in his architectonic harmony). From this discussion it follows that Zaiser's view of harmony as reducibility is just one instance of harmony, not the only one.

Although Kepler's description of harmony is vague, his general position is quite clear. God is not just a dry mathematician, not just a dry logician, but is also a lover of aesthetic beauty. God is a musician. In creating the universe, God used not only the laws of geometry but also those of harmony, particularly the laws of musical harmony. The different things in the world have been arranged, not at random, but according to definite proportions that are pleasing to mind and heart. The different parts of the universe are related to each other by definite harmonic ratios. Hence any attempt to unravel the laws of nature has to take into account the harmonic relations. In the HM Kepler applied the considerations of musical harmony to the study of the structure and laws of nature. His investigations revealed that the ratios of the angular velocities of the planets at the extremities of their orbits were the basic musical intervals: 4/5 (major third) for Saturn, 5/6 (minor third) for Jupiter, 2/3 (perfect fifth) for Mars, 15/16 (major semitone) for Earth, 24/25 (minor semitone) for Venus, and 5/12 (octave plus a minor third) for Mercury. These ratios give the music the cosmic creator sang and still sings. Kepler believed that this discovery was an achievement of utmost importance. For him these results were no mere speculation. Indeed, for him the HM was a valuable book of cosmology wherein he had been able to uncover how the universe existed in its innermost bosom ("qualis existat penitissimo sinu"). It provided him with the first genuine model of the universe ("prima universitatis exempla genuina.")[78] He believed that essential ideas with regard to the origin and nature of the universe, ideas most valuable for a correct understanding of the universe, were unlocked as a result of these studies. He was convinced that this assisted him greatly in his goal to read the mind and plan of God.

However, there were several problems. His mode of thinking and developing ideas in the HM is quite strange and very different from what we are used to. In the case of musical instruments, the length of the vibrating string and the frequency or number of vibrations are the relevant factors for harmony. Corresponding to them in astronomy we should consider distance of planets and their velocity. But in Kepler's work the harmony discovered is neither between the distances nor between the orbital velocities of the planets. Rather, it is only between their angular velocities with respect to the sun. Also the velocities considered are only at the aphelion and perihelion of the planetary orbits. How could Kepler have jumped to a momentous conclusion concerning cosmic harmony under such restricted and limited conditions? How could he have exulted so much at this discovery? The findings seem incomplete, to say the least. However, he could justify himself on the basis of his belief that the universe is a harmonious, unified whole. For if the universe is closely interconnected, then what is true for a part of it must be true for other parts as well.

There were also problems with observational agreement. For instance, Kepler found that the agreement between theory and observational results in the case of Jupiter and Mars was poor.[79] However, here he had a ready explanation. He reminded the readers that in the polyhedral theory, where the position was considered, it was Jupiter and Mars that had the best fit. It should come as no surprise, then, that we get a poor fit for the same planets when we consider motion. In his view for a particular case if we get the best result with respect to the position of the planet, then we should expect to get the worst result for the same case with respect to motion. In other words, we cannot expect the best accuracy in both position and motion (velocity); they are complementary to each other. Undoubtedly, here one is reminded of Heisenberg's uncertainty principle, although, despite his familiarity with Kepler's writings, he does not seem to have noticed this idea about complementarity in Kepler. Surprisingly strange and ad hoc though Kepler's justification may seem, his explanation is not so weak. If he is correct, the poorest result in the polyhedral theory must be the most accurate in the harmonic theory, and, sure enough, such is the case. Saturn–Jupiter gave the poorest agreement with observation in the geometrical theory, whereas they showed the best agreement in the harmonic theory.[80] Clearly, Kepler was neither a loose nor a careless thinker. And his idea also argues for the complementary nature of geometry and harmony.

The need for taking principles of harmony into account in the investigations of the laws of nature was a direct result of Kepler's emphasis on empirical inquiry. In MC he was quite certain that geometry would unlock the secrets of nature. Thus although he did notice discrepancies between his theory and the observed data available to him, at first he tried to explain them away. However, upon coming to know about Tycho Brahe's accurate observations, he wanted to check his theory with the Dane's treasure of observations. This comparison with observational data revealed that his theory based on geometry was not fully accurate, which brought home to him the notion that geometry alone could not discover the laws of nature. He found that when he took into consideration the harmonic laws also, he could get the correct results. Both geometry and harmony had to be brought in if a complete answer was sought.

The incompleteness of geometrical considerations and the consequent need to introduce harmonic principles Kepler explained in the *Epitome*. According to him, "The archetype of the mobile universe is formed not only from the five regular solids from which the planetary paths and the number of motions would be fixed; but is also formed by the harmonic proportions of the six parts in agreement with which the motions themselves have been tuned to the idea of a certain celestial music, or of a six-part harmonic chord."[81] He expressed the complementary[82] nature of geometry and harmony in HM as well: "They [the harmonies] provided, so to speak, nose, eyes, and other members to the statue, whereas the latter [the regular solids] prescribed only the rough external quantity of its mass."[83] He argued that the fact that geometry alone was inadequate and hence the harmonies had to be invoked, far from diminishing the beauty of the universe, enriched it and rendered it more complete.

For this reason, in the same way that it is not usual for the bodies of animated beings, or the volumes (blocks) of stones to be shaped in conformity with the absolute standard of some geometric form, but that they should deviate from the external spherical shape, however elegant it might be (nevertheless, retaining the exact measure of its volume), in order that the body may acquire the organs necessary to life and that the stone may receive the image of the animated being; so, the proportions which should be prescribed for the planetary spheres by the solid figures, were less [in value], and as they affected only the body and its material substance, they ought to yield to the harmonies as much as necessary in order to make them come closer and [contribute] to the beauty of the motions of the spheres. . . .[84]

According to this passage, geometry cannot take a supreme and unyielding position but must make concessions to harmony, insofar as they are necessary to bring about the best and most beautiful universe.

This principle had its reverberations in his religious thinking as well. Earlier, God was only a geometer, but now God has become a musician also. The initial order God placed at creation was such that the motion of all the planets together gave out a harmonious polyphony.

Universalizability (Generalizability)

The principle of universalizability states that the results of rational inquiry, if true, are applicable not just to a few, individual instances but to all instances of the same kind. In fact, in Kepler's view this general applicability is a criterion of truth in the sense that the wider the applicability of a result or principle, the greater its probability of being true. AN makes this point explicit when it argues that whatever is false in a hypothesis is peculiar to it and can be absent. On the other hand, general applicabilty can render a hypothesis true.[85] The laws of nature are really true because they are universally applicable. According to him, this principle can weed out the accidental laws from the true ones, since accidental laws will not pass the generalizability test. Epistemologically speaking, this principle was the basis of many of the inductive generalizations Kepler made during his scientific research. For instance, he did extensive research only on Mars, but he applied the results to all the planets. He was convinced that his findings on Mars were true, hence they must be generalizable or universalizable and so applicable to all the planets.

The principle of generalizability along with the principle of unity had a decisive role to play in his celestial physics. Mechanics in his time was a relatively developed branch of science. Generalizability guaranteed him that the true principles of terrestrial mechanics must be applicable to other parts of the universe, including the heavens.

He did not hesitate to apply the laws of mechanics to the motions of celestial bodies. He applied the principle on many occasions in his investigation of the motion of planets.

Kepler's adherence to this principle also explains the exuberant exaltation he expressed when he had made his discovery of the cosmic music in the HM. He could rejoice so exceedingly because he had made a discovery applicable to the whole cosmos, he had found the most universal fact, and hence the most true fact.

Economy of Nature

The principle of economy of nature holds that nature is economic, deriving the most from the least and maximizing its resources. "Nature does nothing in vain," "Nature uses the least amount of resources,"[86] Kepler asserted time and again. For him nature is the paradigm of an efficient machine. In epistemology this principle manifests itself in the idea that a good explanation uses the least number of explanatory factors.

This principle was deeply rooted in his idea of God. For if nature has been created in accordance with the blueprint in the mind of God, the wisest and best creator, then it has to act in the most intelligent and cleverest way possible. To let nature waste any of its resources, or even to have a tendency to do so, is not worthy of the best creator.

We see Kepler using this principle of economy frequently in his astronomical work. For instance, one of his main criticisms of Tycho's and Ptolemy's systems was precisely that "the motions were unreasonably [frustra] multiplied by Brahe as they were by Ptolemy before him."[87] He used this principle to argue that in the Copernican arrangement of the planets no sphere touched the other and that the intervening space contained no matter except the air of heaven. To posit that this space was filled with matter would be absurd, he insisted. Addressing his opponents, he remarked: "What extravagance of Nature is this, so out of place, so pointless, so little like herself?"[88] Later on, when he was in a dilemma about the source of motion in the solar system, the principle of economy came to his rescue, as will be discussed in part 2.

Simplicity of Nature

According to the principle of simplicity, nature loves simplicity and avoids complexity. A true theory of nature would also have to avoid complexity. The principles of simplicity and economy are closely related; sometimes they are almost indistinguishable. However, the two are different, since, in general, what is most simple is not the most economic and vice versa. The most simple way, at least conceptually, to traverse a height is to take the shortest straight-line path from the lowest to the highest point. But this is not the most economic, as engineers know well from planning roads to the top of a mountain. In epistemology the principle appears as the simplicity criterion of truth, which says that, everything else being equal, the simpler a theory is, the more true it is.

Simplicity is a vague concept since it can appear in so many different shades and forms. Kepler referred to at least two kinds: systematic and structural. Systematic simplicity can be defined as the ability to account for a maximum number of varied phenomena by means of a minimum number of postulates. If there are two theories, both capable of explaining the same number of phenomena, then the one that uses fewer postulates will be considered the simpler. Relativistic mechanics, for instance, is considered simpler than its rival, classical mechanics, because relativistic mechanics can account for more phenomena with fewer postulates and assumptions.

Structural simplicity refers to what the different components of a thing are and how they are arranged. The fewer the number of components, the less the degree of complexity in the organization of the components, the more simple the thing will be. A theory that has to resort to epicycles to explain phenomena is less simple compared to a theory that can explain the same phenomena equally well without them. A theory that can reduce the number of epicycles needed to account for phenomena will be simpler than another that resists such a reduction. Structural simplicity abhors complex structures. This regard for structural simplicity can explain, at least partially, the great predilection people had for circular motion. Kepler rejected the epicycles of Copernicus and the complex arrangement of Tycho because he was convinced that "simplicity is more in agreement with nature."[89]

The twelve principles that I have discussed in this chapter—realism, mathematizability, precision, causality, concomitant variation, dynamism, order, unity, harmony, universalizability, economy, and simplicity—constitute the backbone of Kepler's philosophical worldview. He constantly appealed to them during his long and arduous study of Mars. They guided him along all the way. They confirmed his intuitions. They assured him of success.

3

KEPLER'S SCIENTIFIC IDEAS

Among the founders of modern science Kepler occupies an immortal place. But he would not have earned this privileged position for himself had he stopped at his religious and philosophical views, insightful and creative though they were. He could go beyond most of his contemporaries and rise above their world infested with mysterious and occult forces, precisely because he demanded that his system should include empirical or scientific thought as well. He enjoyed speculating, he could soar to the heights of speculative ideas, yet he never lost contact with the ground. Time and again he insisted that no idea, however startling and attractive, was any good if it could not agree with experience. This requirement was strikingly evident in his view on astronomy. He declared to Brengger that he was giving "a physics of the heavens in place of a theology of the heavens or a metaphysics of Aristotle."[1] For many Aristotelians, cosmology or study of the universe, especially of the heavenly bodies, was a part of metaphysics. Kepler wanted to give a new direction and scope to the study of the heavens. His was a new science, a new physics in which "I teach a new mathematics of computing not from circles but from natural faculties and from the magnetic properties."[2] As Small and Koyré point out, the emphasis on physical explanation distinguished Kepler from practically all his predecessors and contemporaries (except Galileo). Emphasis on physical interpretation was perhaps the most crucial part of his scientific view. Since demand for physical explanation has become commonplace for us, indeed it has become part of our commonsense wisdom, we shall never be able to appreciate fully how novel and revolutionary were Kepler's contributions to science. Apart from stressing realism, his scientific outlook took in several other vistas as well. For instance, he wrote in the *Epitome*: "I build my whole astronomy on Copernicus's hypothesis concerning the world, upon the observations of T. Brahe, and lastly upon the Englishman, W. Gilbert's philosophy of magnetism."[3]

OBSERVATION

Kepler considered observations uniquely important in his system. Galileo, to a great extent, and Newton, to some extent,[4] failed to give due recognition to this fact. On the other hand, Small and Dreyer seem to have overemphasized this dimension of Kepler's investigations. In this context Caspar has the following to say: "As daring and rich in fantasy as he was in his speculations about the universe, just as thoroughly did he now proceed, taking no step without gathering authorization and confirmation from the observations. Indeed, while following his Mars researches, one almost gets the impression that sometimes he deals with individual tasks and proofs out of pure delight and pleasure in the observations."[5] In this emphasis on observation Kepler departed decisively from his predecessors and contemporaries alike (except for Galileo). Indeed, others considered observation important, but neither in the way nor to the extent that Kepler did.

Plato's position on the importance of observation in the acquisition of scientific knowledge is very controversial. Most of his writings, especially the earlier ones, leave little doubt that observations played no major role in his science. For instance, in the *Republic* he had the following to say:

Thus we must pursue astronomy in the same way as geometry, dealing with its fundamental questions. But what is seen in the heavens must be ignored if we truly want to have our share in astronomy. . . . Although celestial phenomena must be regarded as the most beautiful and perfect of that which exists in the visible world (since they are formed of something visible), we must, nevertheless, consider them as far inferior to the true, that is, to the motions . . . really existing behind them. This can be seen by reason and thought, but not perceived with the eyes.[6]

According to this passage, astronomy or science should engage in the study of the fundamental questions and that is to be done not so much by investigating the fleeting and inferior observed phenomena as by discovering the real motion behind the observed ones. Such fundamental questions can be tackled only by the eyes of reason, not of vision. Just as in geometry the power of reason is decisive, so in astronomy the same power should predominate. According to the second sentence of the passage above, we must reject what is seen externally, which, according to the third, must be considered inferior. Plato himself seemed to be hesitating between ruling out and undervaluing the role of observation in empirical science. In his later writings he seemed to attach some importance to observations. In the *Timaeus* he wrote: "Sight, then, in my judgment is the cause of the highest benefits to us in that no word of our present discourse about the universe could ever have been spoken, had we never seen stars, sun, and sky. But as it is, the sight of day and night . . . has caused the invention of number and bestowed on us the notion of time and the study of the nature of the world. . . ."[7] These statements and others like them show that observations did have a place in Plato's study of nature; his exact position on the role of observation remains unclear. But whatever be the final interpretation, Plato did not consider observation crucial for science. The same point of view was upheld by the later Platonists as well.

Aristotle, on the other hand, stressed sense observation for the acquisition of knowledge, as is evident from his idea that nothing can be in the intellect unless it enters through the senses first. Consequently, sense observation is a necessary condition for cognition. Aristotle's insistence on observation can be seen from the fact that he is often considered the founder of zoology because of his book *The Natural History of Animals*, where he classified more than five hundred different species. In the study of cosmology, too, he gave evidence from everyday experience. For instance, he argued from experience to establish that the earth has a spherical shape. Even though reliance or insistence on observation was not absent in the writings of Aristotle and his later followers, the experimental probing of nature certainly was not a crucial factor. Furthermore, where experiments were performed (e.g., Jordanus and the opticians), attention to quantitative precision was minimal. The concern remained very much on the qualitative.

Although like Plato, Kepler emphasized the importance of geometry and geometrical method, and although like Aristotle, Kepler made constant use of philosophical principles in his scientific work, he considered precise and quantitative observations crucial for scientific work. He affirmed: "Without appropriate experiments I conclude nothing."[8] Again, he told Fabricius: "Hypothesis must be built upon and confirmed by observations. . . ."[9] In AN he called them *fidissimi duces*, "most reliable guides."[10] Indeed, they were the most reliable guides in his adventurous and challenging search for the secrets of nature. He could trust the verdict of observation, and, once his theory had been attested to by observation, he could rest assured that he had either reached his goal or was on the right track. When his theory was attacked by others, his major defense consisted in referring his opponent to observational data. He asked: "Is there anything in astronomy more certain than the observations either of the sun or Mars? I make judgments from these."[11]

He himself made many observations,[12] despite his own visual problems[13] and the lack of accurate and advanced instruments. He refused to believe the Copernican explanation for the failure to observe stellar parallax. According to Kepler, the universe is not infinite and the stars are not at an infinite distance from us. To verify this claim, he looked for stellar parallax, but with no success. This failure led him to conclude that the parallax of the Polestar must be smaller than eight minutes "because my instrument cannot measure angles smaller than this."[14] In his letters to various people he talked about the many observations he made. For instance, he had observed Halley's comet. Again, in his letter to Galileo he requested the Italian to make some specific observations.[15] Kepler wanted to make observations as accurate as possible; if experimental conditions did not permit accurate results, he wanted to take several observations because "all doubts are to be excluded by many observations."[16]

It seems that he sensed a kind of divine aura surrounding observations. The expressions he used, the respect he showed to them, evoked the presence of something holy and divine. His words about Tycho's observations illustrate this point: "Let all be silent, and give heed to Tycho Brahe, the Dane, who now for the thirty-fifth year is devoting himself to observation, who sees more with his eyes than many others with mental vision, a single instrument of whom cannot be counterbalanced by my whole thought and being. . . ."[17] Obviously this statement involves a lot of exaggeration. At the same time we notice a deep feeling of reverence as though for something divine. He expressed this idea once again in his letter to Maestlin on March 5, 1605: "I consider as an honor only one thing that by divine disposition I have been permitted to be in close contact with Tycho's observations."[18] His association of observational data with divinity was made clear also when he refused to accept an error of 8 minutes because he believed that Tycho's observations were a gift from God and hence deserved to be given utmost importance. Observations overwhelmed him. He wrote to Herwart: "I would have concluded my research on the harmonies of the world long ago, if Tycho's astronomy had not won me over so strongly that I almost went out of my mind."[19]

How much he valued accurate observations is again manifested by his willingness to make exceptions to many other principles he otherwise so zealously cherished in life. He was deeply attached to the principle of circularity according to which heavenly bodies always move in circular orbits. He gave it up, however, when observations convinced him that the Martian orbit deviated from the circle. Similarly, he was convinced of the supreme power of geometry to reveal the structure of the universe, yet when Tycho's observations disclosed that geometrical considerations alone were inadequate to reveal the structure of the universe, Kepler was willing to accept the verdict of empirical data and look for additional laws required to give a complete picture of the universe. He had insisted on individual freedom, he publicly professed selflessness in his scientific work, and he always loved peace and shunned occasions of petty fights and quarrels. But when Tycho's observations were at stake he seemed to have forgotten all these other fine rules that he had striven so much to cultivate. Kepler, the great advocate of personal freedom and selflessness in scientific work, asserted: "This is my opinion about Tycho: although he abounds in riches, like very many rich persons, he does not know how to make proper use of them. Therefore . . .

let us take measures to wring [*extorqueamus*] his treasures from him so that we can improve on them. . . ."[20] Kepler did not hesitate to advocate the use of some force to get the Dane's observations from him!

Why did Kepler insist on observation so much? He was too original to be overwhelmingly influenced by other individuals. Nevertheless, Copernicus and Tycho exerted major influence on him. And of course, his beloved teacher Maestlin also was another important authority, yet in a different way. The success of the Copernican theory, which had a strong observational basis, further reinforced Kepler's belief that a true theory should be based on observations.

In addition, the ideas and works of Tycho convinced Kepler of the unique importance of observations in scientific investigation. He had sent a copy of his MC to Tycho who promptly expressed his deep and personal appreciation for the ingenious book. The Dane agreed with our astronomer's idea that in the universe existed a certain symmetry and harmony, but Tycho made the following point: "However, the harmony and proportion of this arrangement must be [sought for] a posteriori, where the motions and the circumstances of the motions have been definitely established, and must not be determined a priori as you and Maestlin would do; even there they are difficult to find."[21] In his letter to Maestlin on April 21, 1598, Tycho was less inhibited by any constraints of tactfulness: "If improvement in astronomy must be made a priori, by means of these regular solids, rather than a posteriori, we shall assuredly have to wait too long, if not forever and in vain, until someone does it."[22] He did not believe that an a priori method would lead astronomy very far. Such a program was doomed to failure.

However, we must not conclude that Kepler would not have emphasized observation had he not come under the influence of Tycho. Some scholars seem to imply that interest in empirical inquiry in general, and observational data in particular, was a temporary phase for Kepler and was confined to the AN. In their view, his interest was just a temporary deviation from his lifelong career of speculation and "mystical" contemplation. Nothing can be further from the truth. Even in his MC, his highly speculative book, he took observational evidence into consideration. In fact, after proposing his polyhedral hypothesis, he subjected it to empirical verification, checking it with Copernican data. Although there was fairly satisfactory agreement between theory and observation, there were discrepancies; he took special pains to account for these discrepancies.[23]

Kepler stressed the need for observation all through his career because he was a firm believer of the principle of realism. His realism looked for tangible and observable evidence for scientific theories.

INFERENCE TO GENERALIZATION

Kepler's emphasis on observation in no way made him a positivist who refuses to go beyond the observable. He believed that many things, though unobservable, could be known by inference. In fact, many significant results in his science he arrived at by inference, rather than by direct observation. For instance, in his research on the planetary motion he studied in detail only the planet Mars. However, he generalized his finding to all the planets. Thus already in the AN in the beginning of part 4 he wrote: "Whatever I have demonstrated in the third part are applicable to all the planets. Hence without hesitation they can be considered the key to the genuine astronomy. . . ."[24] In the *Epitome* he made this generalization absolutely explicit.[25] Obviously, this generalization in the AN was not based on direct observations, but he had to rely on inference. One may object to this position by saying that in the *Epitome* Kepler gave the eccentricity and position of the aphelion for Mercury and Venus; for Mercury the eccentricity was 0.210 and aphelion at 255 degrees, for Venus 0.00604 and 302 degrees respectively. The objection is not serious. And these determinations came much later than AN. Again, being able to determine the eccentricities and positions of two or three planets does not entitle one to claim that the planetary laws are applicable to all planets. To make such a general claim, one needs to know all the relevant aspects—not just one or two—to which the laws are applied.

Another remarkable instance in which Kepler made an inference to generalization was in his study of starglobes. He argued that the globes of the stars were similar to the earth although "experience is silent about it, since nobody has been there. Therefore experience neither denies nor confirms [the claim]."[26] He stated that he arrived at this general conclusion by inference on the basis of the similarity between the earth and the heavenly bodies.

He refused to believe that something did not exist just because it had not been observed. For instance, he argued that just because a particular comet could not be seen, we should not deny its existence. "You believe that one should see them [comets] if they exist. I deny this. For if they traverse a path far away from the earth, it is not necessary, if they are small, that one be able to see them."[27] In such cases, however, he cautioned that we should only talk of probable existence.

Kepler could make many generalizations because he believed in the principle of generalizability. Obviously, this philosophical principle enabled him to make extremely important generalizations that were crucial for his scientific work.

INFERENCE TO THE UNOBSERVABLE

In the previous section we have been discussing instances of inference to unobserved, but in principle observable, entities. Kepler believed that such inference was applicable to unobservable entities as well. These directly unobservable entities in present-day philosophy of science we call "theoretical entities." His stressing observation in no way ruled out the possibility and reality of his considering theoretical entities.[28] True, he never called them "theoretical entities," but he talked about "forces," "solar emanation," "magnetism," "animal forces," etc., entities that admittedly were unobservable and yet real for him. Unobservable themselves, their effects could be observed and experienced.

Kepler's belief that inference to the unobserved is legitimate was firmly rooted in the principles of causality, realism, and unity. The principle of causality stipulates that for every effect there must be a cause; invisibility or unobservability of the cause cannot mean impossibility of the cause. Along with causality, adherence to realism makes possible the attribution of reality to the unobserved or unobservable world because the effects observed are real and hence the source responsible for them must also be real. Further, the principle of unity holds that the observable and the unobservable realms are interrelated and constitute one unified whole.

Kepler's religious beliefs made him readily willing to make inference to the unobserved and the unobservable and to affirm them as real and existing. In religion, many fundamental claims he accepted as true and existing though unobservable, for instance, the Holy Trinity—what is real need not in every case be observable and such instances must have encouraged him to infer to the unobservable without undue hesitation.

FORCE

One of Kepler's major contributions was that he emphasized the need for moving away from explanation in terms of abstract principles like "substantial forms" and "natural places" to explanation using real, physical forces. He made this point clear when he contrasted his method with Fabricius's. Kepler wrote on August 1, 1607: "The difference is to be found in the fact that you [Fabricius] use circles and I use corporeal forces."[29] This statement implies that in order to explain certain natural phenomena, his friend took recourse to certain metaphysical presuppositions, whereas Kepler relied on real and corporeal forces. For instance, according to Fabricius and other traditional astronomers, heavenly bodies have to move with uniform circular motion because the sky is spherical and its sphericity demands that the motion be uniform and circular. In contrast to this metaphysical explanation, Kepler wanted to consider the forces involved in the motion and their interaction with the planetary body. We shall see that Keplerian force operated midway between the Newtonian force and the traditional Aristotelian contact force without being either one of them.

Kepler's idea of force underwent a gradual but remarkable change during the course of his career: a transition from the souls to real, physical forces as the cause of planetary motion. That the varying and unpredictable souls of the early days gave way to invariant and predictable forces implied a shift from explaining planetary motion in terms of forms to accounting for it in terms of forces. In turn, the explanatory factor shifted from *a definite way of acting* to *something that acts in a definite way*. This major development in his idea of force is most conspicuously brought out in the two editions of MC. In the first edition he believed that the cause of planetary motion was a soul indwelling in the sun, the effect of the power of which diminished with distance. In the AN he discarded the idea of a soul and postulated an immaterial *species* emanating from the body of the sun.

This immaterial *species* had a number of particular characteristics that prevented it from being cast into any known traditional category. Though immaterial, it was not purely spiritual. It was, in a sense, corporeal because it resided in the material body. Kepler clearly stated that the emanation or *species* "flowed from its [the sun's] body out to its distance without the passing of any time...."[30] The *species* was therefore independent of time in the sense that it moved without the passage of time. It lacked any weight.[31] He insisted that, although it was immaterial and weightless, it did not have infinite velocity.[32] Again, it was quantifiable and hence amenable to mathematical treatment. As he put it, "For we see these motions taking place in space and time and this virtue emanating from its source and diffusing through the spaces of the universe, which are all mathematical [geometrical] realities. From this it follows that this virtue is subject also to other mathematical laws [*necessitatibus*]."[33] It follows, therefore, that the Keplerian force described in the AN is an immaterial but corporeal *species* that is seated within a material body and is quantifiable. Thus he had taken a giant step away from the soul of the MC of 1596. Kepler made sure that his readers captured this evolution of his idea when he wrote the important note in the second edition of the MC:

> If for the word "soul" you substitute the word "force," you have the very same principle on which the celestial physics is established in the *Commentaries on Mars*, and elaborated in Book IV of the *Epitome of Astronomy*. For once I believed that the cause which moves the planets was precisely a soul, as I was of course imbued with the doctrine of J. C. Scaliger on moving intelligences. But when I pondered that this moving cause grows weaker with distance, and that the sun's light grows thinner with distance from the sun, from that I concluded, that this force is something corporeal, that is, an emanation which a body emits, but an immaterial one.[34]

According to Max Jammer, this move from soul to Keplerian force "announces the birth of the Newtonian concept of force."[35] Koestler believes that Kepler's was "the first serious attempt to explain the mechanism of the solar system in terms of physical forces."[36]

Kepler's idea of force was midway between the Newtonian and the Aristotelian concepts. With its corporeal and quantifiable nature Kepler's force certainly was non-Aristotelian. It also differed from the Newtonian notion of force in several respects. First, it was a tangential, pushing force propagated from the body of the sun in straight lines producing velocity, whereas the Newtonian idea was of central force producing acceleration; he believed that the body of the sun along with the immaterial *species* rotated continuously, thereby whirling the planets around the sun. Furthermore, the Newtonian force could act on bodies at a distance without material contact, whereas Kepler still subscribed to the Aristotelian idea of contact forces.

This idea involved serious philosophical problems. For instance, if the emanation was immaterial, how could it interact with material bodies, causing them to move? To this question Kepler replied that though immaterial, the force was directed to matter and exerted no influence in the space between the sun and the planets. The force was not like odor that gets spent up in the intervening space. "It is propagated through the universe . . . but it is nowhere received except where there is a moveable body. . . . [Like light] it has no present existence in the space between the source and the object which it lights up, although it has passed through that space in the past; it 'is' not, it 'was,' so to speak."[37] Again, Kepler said that the immaterial emanation could affect material bodies, which seems to bring up the problem of the interaction between body and soul. His only answer to the challenge was to point his critic to the analogy of light, in which, he argued, such an interaction took place. However, we know that this is only an analogy because there are marked differences between the action of light and of the immaterial emanation. For instance, in an eclipse, light is blocked off and the action of light stops, whereas the action of the immaterial emanation continues to be effective, since planets continue to move at the time of eclipse. The argument from analogy, even if it is valid, cannot solve the philosophical problem but only minimize the strangeness of the case. After all, one can ask how immaterial light can affect material bodies; the basic problem remains unaccounted for.

Kepler indeed seemed to be quite uneasy about the ontology of this emanation. He seemed to have great difficulty explaining exactly what the nature of this force was, although he had hardly any difficulty describing what it did. Comparing the emanation with magnetism also could not get him out of difficulties because with magnetism also the force had serious differences. Magnetic force is attractive or repulsive, whereas this one is tangential. Such a magnetic force would change the orbit of planets considerably. On the other hand, the planets would have to continue moving unaltered in their orbits.

The Keplerian idea of force had its roots in his philosophical and religious ideas.[38] The principle of dynamism, based on a dynamic God, states that our universe is dynamic. The presence of forces is an external manifestation of this dynamism because force is the cause of motion and other activities. The principle of causality requires that the activities in the universe should have an active cause: planetary motion should have an active cause, and Kepler postulated force as the efficient cause of planetary motion.

His all-important concept of dynamic force had its firm basis in his religious view, especially his idea of the trinitarian God. In fact, force was the reflection of the Holy Spirit. He argued that just as God the Father acts through the Holy Spirit, the sun diffused and bestowed motive power across the *intermedium*. Then he affirmed that his description was not a mere heuristic analogy but the true picture of how the universe existed and operated.[39]

MASS

The development of the concept of mass was another important contribution of Kepler. The concepts of mass and force are closely related, complementary concepts, as Max Jammer points out. This complementarity becomes obvious when we realize that force refers to the cause of motion, whereas (inertial) mass to what resists motion. Jammer presents another insight: the idea of force had its origin in the Aristotelian concept of form (the soul being the form of the living being), whereas mass took its origin from the Aristotelian concept of prime matter. Since substantial form and prime matter are complementary facets of the same reality, force and mass should also have a similar relation.

Kepler proposed his idea of mass in AN in connection with planetary motion under the influence of the force from the sun. He said that when the sun's force tried to take the planet along with it, the planet resisted such a motion, since as a material body it had an inherent property "to rest or to privation of motion."[40] A clearer description he gave in the *Epitome* when he attributed to the planets "a natural and material resistance or inertia to leaving a place, once occupied."[41] In his note to the second edition of MC he made the point more openly still: "Clearly the bodies of the planets in motion, or in the process of being carried round the sun, are not to be considered as mathematical points, but definitely as material bodies, and with something in the nature of weight (as I have written in my book *On the New Star*), that is, to the extent to which they possess the ability to resist a motion applied externally, in proportion to the bulk of the body and the density of its matter."[42] From this statement can be inferred: (1) (inertial) mass has the nature of weight; (2) it resists a force which is externally applied; (3) this resistance is proportional to the quantity of matter in the body. The Keplerian concept of mass fell short of the Newtonian in certain significant respects. For instance, although Kepler rejected the doctrine of natural places, he adhered to Aristotelian inertia; he still lacked the modern concept of inertia. Again, the resistance offered by the material body was to motion, not, as in Newton's second law, to acceleration.

Despite these drawbacks, Kepler's idea of mass had original features. For example, as in the case of force, mass is quantifiable. Describing the property of mass or *moles*, as he called it, he wrote: "If two stones were placed in any part of the world, near each other yet beyond the sphere of influence [*orbem virtutis*] of a third cognate body, these two stones, like two magnetic bodies, would come together at some intermediate place, each approaching the other through a distance in proportion to the mass of the other."[43] The distance each stone will move depends on the ratio of their respective masses, and, just like the distances, the masses can also be quantified.

Furthermore, like force, mass is something dynamic and active. It actively fights with the force emanating from the sun: "The transporting power of the sun and the impotence of the planet or its material inertia fight with each other [*pugnant inter se*]."[44] A closely related idea is that the time period of the revolution of the planet is dependent on the mass, just as it is on the force from the sun. This idea, too, departed from the past; according to the traditional view, the celestial bodies offered no resistance to motion and so the matter in them had no role in determining their period of revolution. True, Kepler did not arrive at the Newtonian concept of mass, but he definitely moved away from traditional ideas and contributed to the development of the modern concept.

How did he come up with these new ideas? Reflection on the observed phenomena certainly had a crucial role to play in this context. He wrote in the *Epitome*: "If the matter of celestial bodies were not endowed with inertia, something similar to weight, no force would be needed for their movement from their place; the smallest motive force would suffice to impart to them an infinite velocity. Since, however, the periods of planetary revolutions take up definite times, some longer and others shorter, it is clear that matter must have inertia which accounts for these differences."[45] As in the concept of force, the philosophical principles of dynamism of nature and of causality also must have played an important role in the development of Kepler's concept of mass.

PHYSICAL EXPLANATION

The emphasis on physical explanation in the study of nature was another one of Kepler's original contributions. Gingerich points out that "Kepler was the first and until Descartes the only scientist to demand a physical explanation for celestial phenomena."[46] Kepler insisted on the physical as a necessary condition for any interpretation of natural phenomena. Despite its extreme importance in physical science, no fully satisfactory definition of a physical explanation has emerged, nor is it my purpose to develop a formal definition. An explanation can be looked upon as an attempt to unfold (*explico*) or to elucidate how a phenomenon has come about the way it is. An explanation of spectral lines, for instance, consists in pointing out that they are formed by electron jumps from one energy level to another. Notice that such an elucidation involves theoretical terms like "electron," "stationary orbits," "electronic states," "energy levels," and "electron jumps." Whatever one may say about physical explanation, for Kepler, at least, it was an unfolding where the cause of the phenomenon is given in terms of physical, real factors. Such an explanation, he argued, must remove the "cause of wonder" by specifying the real causes.

Kepler's idea of physical explanation is intimately connected to, and a direct application of, the philosophical principles discussed in chapter 2. These philosophical principles provide the rational basis for a physical explanation. For instance, one of the chief requirements of a rationally acceptable explanation is that it should not violate the philosophical principles of simplicity, economy, causality, etc. Undoubtedly, this scientific view of Kepler had its basis in his philosophical view.

What is "physical"? Although Kepler gave no formal definition of "physical," he did present a good idea of what he meant when he discussed his hypothesis of the oval orbit of Mars in a letter to Maestlin on December 22, 1616: "The hypothesis is physical because it uses the physical example of a magnet. It is physical, i.e. natural, because it is true and is derived from the very internal nature of planetary and solar bodies. It is physical because it pertains to all modes of natural motion."[47] This is the most detailed and least unclear description of his idea of the physical that I have been able to find in his voluminous writings. The passage gives three different reasons or criteria for considering a hypothesis H physical: (1) H is physical if it involves real forces (since for Kepler magnetism was a paradigm of a real physical force, this interpretation seems to be correct). (2) H is physical (i.e., natural) if it is deduced from the very internal nature of bodies like the sun and the planets; H is physical if it is real and derived from the internal, as opposed to the accidental or transitory, characteristics of material bodies. (3) H is physical if it accounts for all modes of natural motions of planets: H is physical if it can account for all real observable phenomena under consideration. This interpretation (3) seems to me to be correct because Kepler is talking of H in the context of explaining the different observed motions of Mars. So, Kepler himself set down three conditions that specify the meaning of "physical."

As we have already noted, for a physical explanation he stipulated one more condition: all explanations in science must be consistent with philosophical principles. This point he brought out very clearly when he talked about his aim to mechanize the heavens. He sought not only a mechanical explanation, but an explication where the causes involved were characterized by the principles of simplicity, unity, etc.[48]

The question now is whether these different requirements individually or collectively constitute the sufficient condition for a physical explanation. Kepler, as usual, was not clear about this question. If they collectively constitute the sufficient condition, what is the status of each individually? Is each principle a necessary condition for physical explanation? I think that together they are collectively sufficient and individually each is a necessary condition to unfold a satisfactory physical account of phenomena. For just the presence of physical force alone cannot render an explanation physical; it should be able to account for observed phenomena. Conversely, just being able to account for observed phenomena will not make research findings physical because a metaphysical working-out for such phenomena could suffice, as the Aristotelians always held. The Ptolemaic method could account for the phenomena, but Kepler never accepted that path as physical. Furthermore, just because an explanation is derived from the internal nature of material bodies need not mean it is physical. This condition can guarantee that the explanatory factor has a basis in reality, but not that it can account for the phenomena itself. Again, consistency and coherence are necessary for any acceptable explanation, but having these two qualities need not entitle the presentation of the findings to be a physical explanation—purely mathematical or logical explanation can be both consistent and coherent without being physical. I think that a variety of different factors collectively constitute a physical explanation for Kepler.

Thus a physical explanation, according to Kepler, accounts for or gives reasons for the true working of a phenomenon in terms of matter, known or knowable forces, and other theoretical entities that have a real basis in the internal, as opposed to the transitory or superficial, characteristics of matter, and that does so in a way consistent with certain philosophical principles.

A prime example was his treatment of planetary motion. The explanation was in terms of the effluvia emanating from the whole body of the sun. This emanation was like a magnetic force whose effects could be explained, although its cause remained unknown. The effect of the force could account for the observed motion of the planets around the sun. Finally, this example harmonized with the principles of simplicity, generalizability, and so forth.

The discussion above can reveal the shortcomings of some earlier attempts by Kepler scholars to describe his idea of "physical" and hence of "physical explanation." Holton thinks that mechanical explainability was the heart of Kepler's idea of physical explanation: "The physically real world, which defines the nature of things, is the world of phenomena explainable by mechanical principles."[49] This obviously raises questions about the meaning of "mechanical principles." Holton says that "mechanical" refers to "the world of objects and of their mechanical interactions in the sense which Newton used."[50] In other words, the physically real must be explainable in terms of matter and forces or interaction. However, the Newtonian forces could act at a distance. Kepler, on the other hand, still adhered to the notion of contact forces. He certainly wanted a scientific explanation to be mechanical in the sense that it involved matter and (contact) forces.[51] But he wanted much more. An explanation may be mechanical, but not real. Galileo's explanation of the tides was certainly mechanical but failed to correspond to reality;[52] Kepler did not go along with the Italian's position.[53] Thus explainability in terms of mechanical principles was only a necessary condition for a physical explanation. For Kepler another condition must be met: consistency with his philosophical principles. He was no Baconian empiricist or Laplacian mechanist. Indeed, he believed in empiricism and mechanism, but both his empiricism and mechanism were grounded deeply in his philosophy.

This insistence on physical explanation as a necessary condition for scientific inquiry was one of Kepler's new and original contributions. This originality can be seen from the fact that Aristotle and the Aristotelians in general did not emphasize the physical as a necessary element in their explanation of natural phenomena. Being guided by the principles of the *Posterior Analytics*, very often their interpretations consisted in logical deductions from certain presuppositions assumed to be true of the universe on the basis of logical rather than physical entities. For instance, consider Aristotle's explanation for the observed fact that heavy objects fall to the ground. He explained the phenomenon in terms of "natural places." The "natural place" of an earthy object was the center of the universe and so it would always tend to move towards that point even if no physical object were there. On the other hand, Newton explained the same phenomenon in terms of an attractive force between the earth and the object. Newton introduced a definite physical factor into his explanation. Of course, there were philosophical problems with Newton's forces, which Leibniz and Berkeley were not slow to point out. And even several thirteenth-century scientists did indeed introduce physical factors into their explanation of natural phenomena like the rainbow (I have already pointed out several marked differences between Kepler and them); nevertheless, their physics still remained Aristotelian. Our astronomer, on the other hand, emphasized both physical explanation and a new physics.

Another consideration also can bring out the originality of Kepler's contribution. Often opposition from others, especially from one's own peers in the field, is a reliable gauge to assess the degree of originality of a new view. An original idea often goes against, or at least challenges, the established set of ideas and norms and thus becomes a threat to the status quo. Hence opposition. If such opposition be regarded as a criterion, then Kepler's view was highly original indeed. His emphasis on physical explanation drew opposition from friends and foes alike.

Perhaps the most vehement opposition came from the clergyman Fabricius, Kepler's persistent and tireless correspondent. Fabricius's strong voluntaristic tendencies surfaced in his reaction to Kepler's emphasis on the explanation of natural phenomena in terms of physical causes. According to voluntarism, God's will is responsible for the different events and phenomena in the universe and so their explanation is to be sought in God's will rather than in any physical causes, which are responsible only indirectly (*mediate*), if at all. Fabricius echoed the same sentiment in his reply to Kepler in a letter on February 7, 1603,[54] asserting that causes of sickness and other natural phenomena were to be found in the will of God (*ordinationem divinam*). Taking place in the universe were so many events, none of which depended on physical causes. He illustrated his case with an example. A stone can be a boundary marker, a road sign, or the cornerstone of a house. The stone's function depends, not on its internal nature, but on the desire and will of humans.[55] Looking for the cause in the stone itself for its being sometimes a milestone, sometimes a cornerstone, etc., would be stupid. Fabricius pointed out that even Holy Scripture attested to his position, for according to the Bible, the stars are endowed with light in order that they may be signs—not causal agents to explain physical phenomena. They are just a passive medium through which God's will is manifested. He concluded his attack: "Therefore it is exceedingly gross [*crassus*] and worthless [*carnale*] to wish to refer all the works of God to physical causes or to philosophical standards. We have seen how invalid the physical and philosophical explanations are from many works of God. . . . By just one single little word of God . . . a multitude of arguments can be brought forth [for the solution of an issue] that cannot be solved sufficiently satisfactorily by all philosophical secrets [wisdom]."[56] According to Fabricius, Kepler's search for physical causes was simply wasted effort, totally unworthy of a believer like his friend.

Kepler was far from being impressed with Fabricius's arguments. Being a firm believer in God, Kepler recognized the efficaciousness of God's will and power. But he refused to go along with his friend's naive interpretation of that will. According to Kepler, God may permit certain events. This would imply that the events follow the laws of nature set by God at the time of creation. God does not interfere with them. God permits certain events to take place at certain times, but God does not will them to existence. Fabricius's position would make God a continuous creator, since every event has to come to existence by God's will. For Kepler, God is the continuous conserver of the universe. Furthermore, he pointed out that his friend's position had to resort to explanations purely in terms of occult divine relations. Regarding the will of God, Kepler argued that Fabricius's position, which relegated the events and operations in the universe to the arbitrary will of God, would render rational science impossible.[57]

Kepler pointed out further that in his eagerness to attribute everything to God, Fabricius was making God a part of nature. Our astronomer then characterized his and his friend's positions: "For you God turns to the constellation to give rise to the different temperaments and inclinations. For me the sublunar souls look after the same function."[58] According to Fabricius, to account for many natural phenomena, one had to resort to God's direct intervention, whereas according to Kepler, such an account could be given by means of the sublunar souls created by God. Later on Kepler would exorcise these souls and bring in forces in their place. What he wanted to say was that he could explain many workings of nature in terms of natural forces rather than by God's direct intervention.

Not even Maestlin appreciated this emphasis on physical explanation. Kepler informed Peter Crüger, professor of mathematics in Danzig, that "Maestlin used to laugh about my endeavors to reduce everything, also in regard to the Moon, to physical origin. In fact, this is my delight, the main consolation and pride of my work, that I succeeded in that. . . ."[59] Writing in this context, Maestlin took his disciple to task:

> When you write about the moon, you say that you would like to adduce physical causes for all its inequalities. I simply do not understand this. On the contrary, I believe that physical causes can be dismissed altogether, and that it is fitting to explain astronomical phenomena only through astronomical methods, by means of astronomical, not physical, causes and hypotheses. For calculations demand astronomical bases from geometry and arithmetic, which, so to speak, represent their wings, rather than physical hypothesis, which would more likely confuse than instruct the reader.[60]

This remarkable letter, written in 1616, long after the AN had been published, makes clear how deep rooted the Ptolemaic conception of astronomy was even in the mind of a moderate like Maestlin. Kepler replied that his teacher's objection arose from an ambiguity concerning the concept "physical." He believed that once this notion was clarified, Maestlin's objection would be taken care of; the objection came, not because the physical method was false, but because the concept had not been clarified.

In the marginal notes of AN he tried to clarify the concept "physical." A physical hypothesis must correspond to something real, it cannot be just a geometrical hypothesis based purely on geometrical assumptions. "I shall accept only that which cannot be doubted as truly real and therefore physical. . . ."[61] Secondly, "physical" refers to something proper to the very nature of the body under consideration, as opposed to something accidental. Thus nonuniform motion of planets he considered physical because it is "appropriate for the nature of the planets, and is therefore physical."[62] (These ideas agree with the passages from Kepler that I quoted earlier.) He went on to say that his ideas were in accord with the true nature of things and since Maestlin was a seeker after this true nature, his clarifications should put his teacher's objections to rest.

THE SCIENTIFIC METHOD

Kepler made outstanding contributions also in the development of the method of modern science. However, since he never gave any systematic or detailed discussion of his method, we have to depend mainly on his actual practice and occasional statements on the subject. He often called his approach to science an "a priori method." Maestlin commended his devoted student for espousing this method, which for him was a "frontal" approach, while the a posteriori method was a "backdoor" approach. Recommending the *Mysterium Cosmographicum* to Hafenreffer, the prorector of Tübingen, Maestlin asked in 1596: "For whoever conceived the idea or made such a daring attempt as to demonstrate a priori the number, the order, the magnitude, and the movements of the celestial spheres . . . and to elicit all this from the secret, unfathomable decrees of heaven!"[63] Like his teacher, Kepler also preferred this method, finding it was superior to all other methods. In MC he wrote:

> For what could be said or imagined which would be more remarkable, or more convincing, than that what Copernicus established by observation, from the effects, a posteriori, by a lucky rather than a confident guess, like a blind man, leaning on a stick as he walks (as Rheticus himself used to say) and believed to be the case, all that, I say, is discovered to have been quite correctly established by reasoning derived a priori, from the causes, from the idea of creation?[64]

He was convinced in his early days that this method was the best: "For God knows, this a priori method serves to improve the study of the movements of the celestial bodies."[65]

What is the a priori method? It is a deductive method as opposed to the inductive, a posteriori method advocated by Tycho. Certain hypotheses concerning the explanation of natural phenomena are presupposed.[66] These hypotheses are expressed in mathematical formulations, and particular mathematical deductions are drawn from them. The mathematical formulations and deductions make the approach a priori, since they are considered strictly valid in contrast to the merely empirical observations and generalizations that are involved in induction. The propositions need not always be given a mathematical formulation; however, in his work, Kepler usually did so.

This a priori method seemed to fit well with Kepler's understanding of what astronomy should be.[67] According to him, astronomers should give *causas probabiles* (causes worthy of approval) for the phenomena under consideration. To achieve this he stipulated that the principles of astronomy must be anchored in a higher science, namely physics or metaphysics. The a priori method could provide the *causas probabiles* by way of deduction from higher principles.

I believe that the name "a priori" is misleading and that Kepler's method in essence was the hypothetico-deductive method (H–D method) as understood by several modern philosophers of science. The name "a priori" implies that observation has no real role in the method, which is far from the truth; for despite all the emphasis Kepler placed on metaphysical and mathematical reasoning, despite all the criticism he raised against the a posteriori method of Tycho and others, he also recognized the importance of observation. Thus already in 1595, in the prime of the MC, he wrote to Maestlin: "We can put all our hopes into it [the a priori method] if others cooperate who have observations at their disposal. . . ."[68] To Herwart on July 12, 1600, he affirmed that the a priori speculations must not contradict manifest experience but, rather, agree with it.[69] Therefore the a priori method was not purely mathematical, and still less was it purely speculative. Experience played a significant role. This was quite evident from Kepler's extreme eagerness to see Tycho's observational data in order to verify the findings of his polyhedral theory.

Kepler's a priori method was not purely a priori, rather combined a priori and a posteriori principles. In this regard Kepler's method is very similar to the *argumentatio ex suppositione* that Galileo was to employ years later;[70] this form of argument also has two distinct parts: the a priori and the empirical. The a priori part provides conceptual aids such as axioms and definitions and also specifies that the reasoning involved is to be based on logical dependencies. On the other hand, the empirical dimension relates the propositions obtained by means of a priori arguments to empirical existential statements.

Tycho seemed to have overlooked this point when he criticized Kepler and Maestlin for attempting to build up an astronomy without using accurate empirical data. Kepler insisted on agreement with observed data, and when there was an appreciable discrepancy he looked for an explanation. For instance, he noticed that the polyhedral theory had a great discrepancy: his value for the distance of Mercury was 577 whereas the Copernican data gave 723. He tried to get around this problem by arguing that Mercury was an exception. This planet, he maintained, should not be inscribed within the sphere of the octahedron, as would be expected according to the original model, but within the circle of the octahedron square. This gave a value of 707, much closer to the Copernican value of 723. He tried to give some justification for this special treatment of the innermost planet. The key point of his argument was that the exceptional character of Mercury demanded that it be treated differently.[71] His explanation turned out to be false. But the point is that he took the disagreement with observation seriously and attempted to provide an explanation by using tools available to him.[72]

Kepler's so-called a priori method has one major difference from the H–D method: the hypothesis is not a product of some irrational or purely psychological process.[73] According to him, a scientific hypothesis has a rational basis. This point is very clear from his response to Fabricius: "But you think that I can just invent some elegant hypothesis and pat myself on the back for embellishing it and then finally examine it with respect to observations. But you are far off."[74] Hypothesis is not the figment of one's imagination, as he illustrated by a concrete example: "I demonstrated the oval figure from the observations. . . ."[75]

No doubt he used hypotheses in this sense in the AN. But one may object that in MC and HM such was not the case. For instance, the opponent may say that the polyhedral theory came to him by chance. In his own words, "Eventually by a certain mere accident I chanced to come closer to the actual state of affairs."[76] First of all, Kepler did not use the word "chance" in the sense of "at random" or "irrationally." For him a chance event was simply an unforeseen event, an event that he did not anticipate. Here, for instance, after making the statement above he went on to give the different steps he used to arrive at the hypothesis.[77] Moreover, Kepler truly believed that this hypothesis came to him as a gift from God, the gift of Divine Providence. He continued: "I thought it was by divine intervention that I gained fortuitously what I was never able to obtain by any amount of toil; and I believed that all the more because I had always prayed to God that if Copernicus had told the truth, things should proceed in this way."[78] Here Kepler says that God gave him this hypothesis as a gift. However, this belief did not render the hypothesis any the less rationally founded, because he was convinced that the supremely rational God must have had a good reason for it. Hence even here the hypothesis was not the product of irrationality.

From the hypothesis thus rationally arrived at he made certain rational deductions that he subjected to observational tests. This point also he explained in no uncertain terms: "I put forward [or presuppose] a few things as a groundwork and from there I follow observations."[79] The same idea he expressed in another letter to Fabricius: "It is true that when a hypothesis is constructed from observations and then confirmed by them, I am wonderfully excited afterwards, if I can discover in it something consonant with [*concinnitatem*] nature."[80] This statement specifies that the item derived from the hypothesis is not just anything, but something consonant or in agreement with accepted principles of nature.[81]

Hypothesis formation has at least two constraints. First of all, it must be consonant with his principles. Secondly, it has to have a basis in observation. It is not created in a vacuum, it is not a product of pure imagination. A hypothesis is hypothetical, not because it does not have any basis in principles and observation, but because these bases are not adequate or sufficient.

Kepler's method basically consisted in deriving certain testable deductions from such a hypothesis and subjecting them to observational scrutiny. That this was his general method not only in the AN but even in his highly speculative MC can be seen from a brief examination of his polyhedral theory. This hypothesis had a strong basis in his religious belief in a geometer God. It also took root in his philosophical view that the universe is geometrical, rational, and intelligible to us humans. In a way it had an observational basis too, because, as he narrated, the idea came to him when he inscribed "many triangles, or quasi-triangles, in the same circle, so that the end of one was the beginning of another."[82] He observed that the points at which the sides of the triangles intersected traced out a smaller circle. "The ratio of the circles to each other appeared to the eye almost the same as that between Saturn and Jupiter."[83] Consequently, this hypothesis was not the product of a random process. Once the hypothesis was formed, he deduced from it the relative distances of the planets from the sun. Then he checked the values obtained from the theory against those given by Copernicus.

It can be shown that in all his three major works (i.e., MC, AN, and HM) the method used is basically the same.[84] In all of them the principal goal is the same: scientifically, to discover the structure and operation of the universe; religiously, to give praise and honor to the supreme creator; and philosophically, to establish the universe as rational, intelligible, and accessible to human inquiry. All three use the same kind of tools of inquiry. Sometimes a particular tool may be used more than at another time. For instance, reliance on observational data is more conspicuous in AN than in HM and MC, harmonic considerations are more pronounced in HM than in the other two works, geometrical inquiry is more conspicuous in MC. In all the three works, however, he basically employs the H–D method in the sense we have discussed above—in AN as has been discussed by many Kepler scholars, and in MC and HM as we are about to see. In MC the basic contention is the polyhedral hypothesis that the solar system has been created as a series of spheres in which are nested the five regular solids according to a certain pattern. In HM the basic hypothesis is that the whole universe is governed by the rules of harmony, especially those of musical harmony, and that the creator made use of these rules in the creation of the universe. Both hypotheses have empirical backing. From both hypotheses particular observable consequences have been deduced. We have already seen what the empirical deductions were in the polyhedral theory. For the harmonic theory he also looked for empirical verification.[85] He found that it can quite well account for the eccentricities of the planetary orbits. Another remarkable point that once more bears eloquent testimony to the fact that Kepler, even in his most speculative moments, remained a scientist at heart is seen in one of the suggestions he made while discussing the harmonic theory: this theory could yield the age of the universe, or at least of the solar system.[86] He deduced from this theory by calculation the number of years needed for two planets to come into consonance. Then he noted that a consonance between three planets could be calculated. Similarly, the number of years required for a consonance among all the planets of the solar system could be obtained from the theory. He identified the beginning of the universe, or at least of the solar system, with the instant of consonance among all the planets. Thus he argued that his theory could give the age of the universe. Although here he did not carry out the actual calculation, he gave directions for it.[87]

I do not deny that there are differences in the methods of AN, MC, and HM, but I think that the differences are only a matter of degree of emphasis, rather than of basic nature. If we are inclined to think that AN is very scientific in method and that MC and HM are unscientific, then perhaps we are prejudiced in favor of the theory that turned out to be correct. Thanks to hindsight. There is, in fact, no radical difference of method. This is quite understandable because Kepler always modified his method to suit the goal he had in mind. Thus in MC and HM he was dealing with wider, cosmological questions such as Why did the creator make the universe as we see and experience it? He wanted to discover cosmic secrets, laws governing the whole cosmos. On the other hand, in AN he was investigating a specific problem: the laws governing the motion of planets, more exactly of just one planet. In all three works he was investigating laws of nature, but I notice a progressive widening of the goal or scope envisaged as we move from AN to MC and HM. Obviously, here I am not considering the chronological ordering of these books but the progression in the domain explored by the theories. In AN the goal is restricted since it confines itself to the motion of a single planet. In MC it is wider since it envisages the whole solar system. Finally in HM it is the widest, embracing the whole cosmos. At the same time I see a progressive deemphasis on the reliance on observations as we move from AN to HM. I suggest that the difference between AN and HM is a difference between science of specific things (e.g., study of nuclear fusion reaction in the laboratory) and science of cosmic and general phenomena (e.g., study of the origin of the universe). They may differ in the accurate, reliable results they can provide but not in their basic methodology.

Our study shows that Kepler made significant contributions in the development of the methodology of modern science. He can be rightly considered the first among the moderns to use the H–D method effectively in science. His other contributions to the development of modern science include his emphasis on the need for observation and for physical explanation in scientific inquiry. All were crucial for the transition of science from the medieval to the modern.

Part 2

The Discovery of the Laws

4

THE ACCEPTANCE OF COPERNICANISM

Accepting the Copernican hypothesis was the essential first step Kepler took on his way to the discovery of the laws of astronomy. Without taking this step he would never have become the father of modern astronomy. Despite its unique importance, it was no easy step to take because the Copernican hypothesis was beset with problems from all angles: empirical, philosophical, and religious. Yet Kepler embraced the hypothesis wholeheartedly. Several ways of explaining his daring move have emerged. I think that none of these accounts is entirely sufficient; each emphasizes only one or a few of the aspects involved in Kepler's acceptance of Copernicanism. A more satisfactory explanation can emerge only when we consider the interlocking of ideas and arguments arising from all three dimensions of his system of thought—from his scientific, philosophical, and religious views.

THE PROBLEM OF KEPLER'S
ACCEPTANCE OF COPERNICANISM

While discussing Kepler's acceptance of the Copernican hypothesis, a number of significant points are worth emphasizing. His commitment was not haphazard but, rather, wholehearted and total. He was so firmly convinced of the hypothesis that he was not afraid to defend it publicly with all the might he could muster, although he was fully aware of the formidable opposition to it. His own words echo these points. Already in 1598, when he was beginning his career, he affirmed: "I to whom the Copernican view has been the most persuasive cannot in conscience propose anything else either in praise of my own ingenuity or for the sake of human popularity."[1] Again, at the peak of career in the *Epitome*, he wrote: "I certainly know that I owe it [the Copernican hypothesis] this duty, that as I have attested it as true in my deepest soul, and as I contemplate its beauty with incredible and ravishing delight, I should also publicly defend it to my readers with all the force at my command."[2] Clearly, the hypothesis struck such deep roots in his mind that it had become virtually irrevocable. Never even once do we find him calling it into question. To be sure, he admitted that it had defects (e.g., the problem of stellar parallax), but he never let them shake his faith in the theory itself.

Against a strong, prevalent instrumentalist tradition which regarded astronomical hypotheses as mere computational devices, Kepler asserted that Copernicus's hypothesis was not a mere computational tool but a true picture of the universe. This instrumentalist tradition was crystallized in Kepler's time in Robert Cardinal Bellarmine's well-known letter of 1615:

> To say that on the supposition that the earth moves and the sun stands still all the appearances are saved better than on the assumption of eccentrics and epicycles, is to say very well—there is no danger in that, and it is sufficient for the mathematician: but to wish to affirm that *in reality* the sun stands still in the center of the world, and that the earth is located in the third heaven and revolves with great velocity about the sun, is a thing in which there is much danger. . . .[3]

Osiander's preface to Copernicus's DR was also, as we see here, an effort to strike a consonant note with the predominant instrumentalist tune of the day:

> For it is proper for an astronomer to establish a record of the motions of the heavens with diligent and skillful observations, and then to think out and construct laws for them, or rather hypotheses,

whatever their nature may be, since the true laws cannot be reached by the use of reason; and from those assumptions the motions can be correctly calculated, both for the future and for the past. Our author has shown himself outstandingly skillful in both these respects. Nor is it necessary that these hypotheses should be true, nor indeed even probable, but it is sufficient if they merely produce calculations which agree with the observations.[4]

Many scholars following the lead of Kepler himself are of the opinion that Osiander did not accurately represent Copernicus's mind in this preface; indeed, Kepler was the champion of those who rushed to condemn Osiander.

These scholars have several reasons to believe that Copernicus in fact subscribed to a realist view. For instance, he was not content with just "saving the phenomena," he wanted also to explain them. In DR he described astronomy as a "discipline which deals with the universe's divine revolutions, the asters' motions, sizes, distances, risings, and settings, as well as the causes of the other phenomena in the sky, and which, in short, explains its whole appearance."[5] According to him, the earth moves in reality. If we do not observe the earth moving, it is not because the earth is stationary but because we make the observations from a moving earth. Refuting the Aristotelians, he wrote in DR: "We regard it as a certainty that the earth, enclosed between poles, is bounded by a spherical surface. Why then do we still hesitate to grant it the motion appropriate by nature to its form rather than attribute a movement to the entire universe, whose limit is unknown and unknowable?"[6] All these points can support the realism of the Canon.

This was no unwavering realism, however. Copernicus left enough room for Osiander and like-minded people to doubt the seriousness of his realism. For instance, in his dedicatory letter to Pope Paul III Copernicus wrote: "So I should like your Holiness to know that I was impelled to consider a *different system of deducing* the motions of the universe's spheres for no other reason than the realization that astronomers do not agree among themselves in their investigations of this subject."[7] Here he himself gives evidence of instrumentalism—he was looking for a "different system of deducing the motions" ("de alia ratione subducendorum motuum sphaerarum mundi").[8] Kuhn translates the phrase, "method of computing the motions,"[9] which is not inaccurate in this context. Whether we translate the phrase as "system of deducing motion" or "method of computing motion," the passage implies that Copernicus took the hypothesis of a moving earth as a tool to account for observed phenomena better, rather than as a statement about reality. Since the fundamental claim of instrumentalism in astronomy is that hypotheses are mere methods or tools for computing the motions of heavenly bodies, Copernicus's comment smacked of a shade of instrumentalism.

This suspicion is reinforced. Later on in the same preface Copernicus declared that he was following the lead of other astronomers before him who "had been granted the freedom to imagine any circles whatever for the purpose of explaining the heavenly phenomena."[10] Indeed, Neugebauer, referring to Osiander's preface, says: "I realize that one is supposed to be disgusted with Osiander's preface which he added to the DR (in keen anticipation of the struggle of the next generations), in which he, in the traditional fashion of the ancients, speaks about mere 'hypotheses' represented by the cinematic models adopted in this work. It is hard for me to imagine how a careful reader could reach a different conclusion."[11] In fact, highly reputed astronomers of the day, like Maestlin, looked at the hypothesis only as a computational tool. One could certainly argue that Copernicus did not say that it was *only* a computational device or method. His statement is also compatible with realism since for a realist a hypothesis can be both a computational device and a representation of reality. Also, one could argue that the Canon of Frauenburg intentionally formulated the statement in a vague way in order to circumvent any possible objections from the papal experts who were ardent supporters of the traditional view. Although this defense is very plausible, the fact remains that Copernicus's formulation casts serious doubt on his commitment to realism, a suspicion that becomes even stronger when one realizes that his system constantly and generously employed the unreal epicycles in its explanation of planetary motion. Hence serious questions arose about the realistic claim of the new hypothesis. Kepler nevertheless accepted the theory as realist.

Since I have argued earlier that Kepler's realism was partly influenced by Copernicanism, one may ask how this could have come about, if Copernicus wavered in his own realism. I find no serious objection here, because we may say that whether he was consistent or not, Kepler could see the realist implications of Copernicanism and accepted it. This interpretation finds support in one of Kepler's criticisms of his master, namely, Copernicus failed to appreciate his own riches.

Furthermore, while considering the adoption of the Copernican hypothesis by a sixteenth- or early seventeenth-century thinker, we need to dispel all modern textbook ideas about the hypothesis at its inception. According to some, 1543, the year of publication of the *De Revolutionibus*, marked a decisive turning point in the development of the physical sciences. Some others go still further to affirm this year as the very starting point of modern science. For these people a new era was ushered in with DR: in place of the old earth-centered worldview, a new sun-centered outlook was installed. Such considerations have led some to call Copernicus the father of modern science or, at least, of modern astronomy.

Recent and contemporary scholarship, however, has taken a far less exuberant view. More and more it is becoming clear that the unique position accorded to Copernicus is not fully justified. The revolutionary Copernicus of some textbooks was more a creation of later historians than of the actual ideas and arguments in the DR. The demythologization of his contribution has been one of the outstanding results of recent researches on his work. As Kuhn, for instance, points out: "Most of the essential elements by which we know the Copernican Revolution—easy and accurate computation of planetary positions, the abolition of epicycles and eccentrics, the dissolution of the spheres, the sun a star, the infinite expansion of the universe—these and many others are not to be found anywhere in Copernicus's work."[12] In fact, Kuhn sees very little revolution in the DR. "In every respect except the earth's motion the DR seems more closely akin to the works of ancient and medieval astronomers and cosmologists than to the writings of the succeeding generations who based their works upon Copernicus's and who made explicit the radical consequences that even its author had not seen in his work."[13]

The thesis of the motion of the earth was not all that novel, either. As Hanson argues, a long line of astronomers and thinkers had held such a view, although they had advocated it for not exactly the same reasons as Copernicus. Pythagoras's idea about the motion of the earth is well known. Down through the centuries several others, like Philolaus, Hiketas, Ecphantus, Aristarchus, Heraclides of Pontus, Capella, Erigena, Bacon, Grosseteste, Scotus, Ockham, Albert of

Saxony, Buridan, Oresme, and Cusa,[14] had accepted the possibility of the earth's motion. In fact, Copernicus himself acknowledged his indebtedness to these thinkers in his dedicatory letter to Pope Paul III.[15]

Hanson and Mittelstrass share an opinion similar to Kuhn's. According to Hanson, "Copernicus might well be viewed as the last of the great orthodox planetary theorists. . . ."[16] A careful study of Copernicus reveals that he proposed the new system not because he disliked the ancient system but because he liked it more; not because he disregarded the Platonic principles[17] of astronomy but because he respected them more. This point was made eminently clear in his attack on Johann Werner, the Nuremberg mathematician who challenged some of the orbs of the ancients:

> It is fitting for us to follow the methods of the ancients strictly and to hold fast to their observations which have been handed down to us like a testament. And to him who thinks that they are not to be entirely trusted in this respect, the gates of our science are certainly closed. He will lie before that gate and spin the dreams of the deranged about the motion of the eighth sphere; and he will get what he deserves for believing that he can lend support to his own hallucinations by slandering the ancients.[18]

Copernicus noticed, rightly so, that the Ptolemaic system was in blatant violation of the ancient principles. As he made clear in his letter to Pope Paul III, the Ptolemaists by their reckless disregard for astronomical principles had given rise to a disfigured monster rather than a well-formed human being. The astronomer wanted to rectify flagrant mistakes of Ptolemy and his followers. Ironically, it seems that far from being a revolutionary, Copernicus was a conservative clamoring for a return to original principles; his goal was to reform Ptolemy rather than to replace him. Similarly, DR was not all that revolutionary: Certainly it had new ideas and presented ideas in a way far more convincing than any in the past; but it also had defects and serious deficiencies. Yet Kepler accepted it.

There is no doubt that Kepler was aware of the defects and serious challenges of the new hypothesis. And he was not satisfied with Copernicus's attempts to meet such challenges. For instance, Kepler remained totally unconvinced of Copernicus's discounting of the unobserved stellar parallax on the plea that the fixed stars were at an enormously large distance from us.

Kepler's acceptance of Copernicanism as the only, true theory of the universe becomes even more puzzling when we realize that he was not naive and, far from being a gullible believer, he was an extremely critical thinker. And this critical spirit was not something he developed late in his life; he had it to a remarkable degree even as a boy of twelve.[19] Nor was he an impressionistic youngster who jumped easily onto the bandwagon of anything novel or different from the ordinary. For instance, when accused of innovations in his religious views, he openly confessed that he was not fond of innovation in any way.[20] His extreme hesitation in parting with the circular orbit gives further clear evidence that he could not be tantalized easily by novelty. Nor was he overwhelmed by the authority of eminent men, though he remained respectful towards them. Pythagoras and Plato he considered as his masters, but he disagreed with both of them. Nicholas of Cusa, Copernicus, Tycho, and Maestlin had great influence on him, yet he differed from them significantly.

The problem before us, therefore, is to explain how such a critical-minded, hard-headed, extremely intelligent person like Kepler could have embraced the new hypothesis not just as a better computational device but as the true representation of the actual world, despite all of the system's known defects and challenges. Before I present my own view, let us pause to discuss the findings of two other Kepler scholars on Kepler's acceptance of Copernicanism.

TWO VIEWS ON
KEPLER'S ACCEPTANCE OF COPERNICANISM

Robert Westman

In his study of the reasons for Kepler's acceptance of Copernicanism, Robert Westman considers the influences of religious, philosophical, and empirical ideas. According to him, religion played no significant role; philosophy had some influence; empirical ideas were the most decisive. In fact, empiricism furnished Kepler with clear evidence that the Copernican order was the necessary and sufficient condition for saving the phenomena.

Although admitting that Kepler stressed the importance and centrality of the sun, Westman does not believe that this emphasis was crucial for the astronomer's acceptance of heliocentrism. No necessary connection or correlation existed between accepting of Copernicanism and assigning a preeminent position to the sun. Many persons like Marsilius Ficinus, Kepler's contemporary Robert Fludd, etc., held that the sun had a most important position in our universe, but they never embraced heliocentrism. According to Westman, "the renewed emphasis upon the sun's dignity in the universe in no way favored one astronomical system over another."[21] He believes that "it was only after astronomers began to understand the sun's importance as a center of motion that they came to interpret (or suddenly perceive) the symbolic sun as justifying their view."[22] Thus in his view the religious significance of the sun was not a motive impelling Kepler to accept heliocentrism. It may, however, have had a post-factum justificatory role. "For only because he grasped the essential properties of Copernicanism could he find in the mystical sun a unique astronomical significance."[23] Religious ideas thus had no real role to play in Kepler's first major step towards the discovery of the laws of astronomy.

On the other hand, the aesthetic and mathematical appeal of heliocentrism, its regularity and simplicity, was of major importance in motivating him to embrace the theory. However, above all it was Maestlin's influence, specifically his work on the comet of 1577, that induced the young Kepler to become a staunch Copernican. Maestlin provided him with the "special argument."

In his careful study of the trajectory of the comet of 1577, Maestlin first used the geocentric model to arrive at a satisfactory solution, but with no success. Finally he came to the Copernican view and formed the hypothesis that the orbit of the comet was coincident with the sphere of Venus. To his great satisfaction, this hypothesis provided him with what he was looking for. As he put it: "Having examined everything in Copernicus's book, I discovered at last a certain orb, in book 6, chapter 2, where the latitudes of Venus are explained and since I found out that its size and revolution corresponded to, and satisfied, the comet's appearances exactly . . . it was then established that the comet chose no other place than the sphere of Venus itself."[24] He argued that such a correct and satisfactory result could not have been obtained if the Copernican theory underlying his investigation were false. This result convinced him of the superiority of the new theory.

According to Westman, to accept the place of Venus on this model was to admit the new order of the inferior planets.[25] "Maestlin had not only introduced Kepler to the basic tenets of the Copernican hypothesis, a theory which possessed an immediate aesthetic appeal for both of them, he had also furnished him with operational evidence that the Copernican order was the necessary and sufficient condition for saving the new appearance, and perhaps for saving *all* celestial phenomena."[26]

At the same time, however, Maestlin looked on the new theory only as a calculational tool, as Westman admits. We know on the other hand that Kepler insisted that Copernicus's view was far more than a calculational tool: the hypothesis represented the true structure of the universe. Despite this objection, Westman continues: "Although he [Maestlin] recognized it as a speculation, such a dramatic demonstration of the potency of the new hypothesis in accounting for the trajectory of this ephemeral phenomenon, where the old theory had failed must have strongly affected the young and impressionable Kepler."[27] Westman seems to say that even though there were significant limitations to Maestlin's investigations, the impressionability of youth had the upper hand and persuaded Kepler to accept the new hypothesis. This explanation, although possible, does not seem fully convincing. For one thing, although Kepler was indeed a daring youth, he was also very independent and critical minded. Besides, if he was carried away by impressionability, he would have given up the hypothesis later on, unless he had compelling reasons for holding onto it. Westman's argument seems inadequate to account for Kepler's wholehearted and continuing acceptance of Copernicanism, especially because Kepler went far beyond Maestlin in taking the new theory as true, not as a mere calculational tool.

Again, the claim that through his study of the trajectory of the comet Maestlin provided Kepler with operational evidence that the Copernican order was the necessary and sufficient condition for saving the new appearance seems questionable. The hypothesis may well have been sufficient, but not necessary. The Copernican model was equivalent to other models, and this was well known to astronomers.[28] For instance, already in 1585 Tycho in his manuscript *De Marte* referred to his model as an "inverted Copernican model,"[29] thus implying that the Tychonic and Copernican models were equivalent. Kepler himself affirmed this equivalence time and again in

AN.[30] Westman admits this objection, but says that it was unlikely that the astronomers comprehended the complete interchangeability or formal equivalence of the two models. This answer is not fully satisfactory. As long as they knew that the two were interchangeable in the relevant respects, they did not have to argue that either system was a necessary condition for saving the appearance. Kepler argued that other hypotheses also could represent the same observations of the comet.[31]

Still more striking is that in the end, Maestlin's idea turned out to be incorrect; Kepler himself referred to this fact in his later writings. Commenting on this point, Westman writes: "Ironically, this hypothesis of a circular, heliocentric cometary orbit—which had been erroneously explained by both Tycho and Maestlin—struck Kepler as a significant counter-instance to the Ptolemaic theory. . . ."[32] This conclusion, if true, would mean that the greatest Copernican of the day, one of the first persons to embrace the new theory publicly and wholeheartedly, did so on the basis of a false explanation of the orbit of a comet.

It seems to me that Westman gives too much significance to Maestlin's study of the comet, which was only one of many factors that induced the young German to accept Copernicanism. The study could not have been the single most important influence, much less could that work have provided Kepler with a necessary and sufficient condition for the veracity and superiority of the new system. And in fact Kepler never mentioned the research on the comet as the crucial reason for his accepting Copernicanism; only rarely did he even mention it at all. Thus in MC chapter 1, he wrote:

> Yet I did not embrace this cause rashly, and without very weighty support of that famous mathematician, my teacher Maestlin. . . . By another particular argument he furnished me with a third reason for accepting the theory when he realized that the comet of the year 1577 moved in complete conformity with the motion of Venus stated by Copernicus, and, by a conjecture drawn from its altitude's being greater than the Moon's, that its whole path was in the actual sphere of Venus.[33]

The only other place I know, where he referred to the role of the comet in his acceptance of Copernicanism was in his letter to Maestlin, written on October 3, 1595: "Therefore I bring forward three reasons, moved by which I have always remained stuck to Copernicus. . . . The third is that Venerian comet of yours."[34] Kepler did not elaborate on this point. Hence Maestlin's work on the comet provided only one reason—and then not the most fundamental or crucial reason—in Kepler's acceptance of the new theory. Kepler never referred to the comet as a reason for adopting Copernicus's hypothesis after the publication of MC in 1596. If it had been fundamental, he would have mentioned it later on when he referred to why he adhered to heliocentrism. In the second edition of MC he appended a note to this case of the comet: nowhere did he say that the comet was crucial for his acceptance of the theory. On the contrary, he made clear that Maestlin himself had admitted that the comet hypothesis was false. According to Westman, Kepler's silence was due to the fact that "the comet has outlived its original usefulness as an argument in support of the new cosmology."[35] This explanation is plausible but not fully satisfactory. The collapse of the argument—if it were indeed "crucial"—would have shaken Kepler's (and others') faith in the new theory, and so he would surely have brought up the issue in his disucssions, especially because they were meant to convince his unbelieving, hostile contemporaries. I think the more likely explanation is that the work on the comet had played only a partial or minor role in his acceptance of Copernicanism, and the eventual discovery of the study's incorrectness had no serious bearing on his over-all commitment to heliocentrism. Westman's arguments need to be supplemented.

E. A. Burtt

In *The Metaphysical Foundations of Modern Science*, E. A. Burtt discusses the question of Kepler's acceptance of Copernicanism.[36] He admits that our astronomer's specific reason for adopting the new theory is at least in part obscure, although there is clear evidence of the influence of philosophical principles such as "Natura simplicitatem amat," "amat illa unitatem," etc. However, "the most potent single factor in his early enthusiasm for Copernicanism appears to be found in its exaltation in dignity and importance of the sun. Founder of exact modern science though he was, Kepler combined with his exact methods and indeed found his motivation for them in certain long discredited superstitions, including what is not unfair to describe as sun worship."[37]

Kepler's outstanding mathematical knowledge and the special mathematical attractiveness of the theory also must have played a significant role in his choice. "But the connection between Kepler, the sun worshipper, and Kepler, the seeker of exact mathematical knowledge of astronomical nature, is very close. It was primarily by such considerations as the deification of the sun and its proper placing at the center of the universe that Kepler in the years of his adolescent fervor and warm imagination was induced to accept the new system."[38]

To substantiate his position Burtt adduces passages from Kepler's 1593 disputation at Tübingen: "In the first place. . . of all the bodies in the universe the most excellent is the sun . . . which alone we should judge worthy of the Most High God, should he be pleased with a material domicile and choose a place in which to dwell with the blessed angels. . . . The sun alone appears, by virtue of his dignity and power, suited for this motive duty and worthy to become the home of God himself, not to say the first mover."[39]

Certainly these remarks highlight the great importance Kepler attached to the sun. But does this evidence warrant Burtt's claim? Do these passages show that "sun worship" was "the most potent single factor," "the main sufficient reason," for Kepler's adoption of Copernicanism? I think not. After all, Copernicus referred to the sun in quite similar terms, but he did not propose his great regard for the sun as the main reason for postulating his theory.

There is no doubt that Kepler had a deep regard for the nobility of the sun, but he never presented it as the most potent single reason for adopting the new hypothesis. Burtt contends, giving just one reference for his claims, that ever since 1593 Kepler always cited the central position of the sun among his expressed reasons for accepting Copernicanism, "usually first." However, in MC of 1596 Kepler did not give this as his first or most important reason; in fact, in the first chapter, where he discussed elaborately why he embraced heliocentrism, he did not mention directly the central position of the sun or its divine symbolism.[40] The centrality of the sun was not one of the three reasons he gave in his letter to Maestlin on October 3, 1595.[41] Of course, other passages mention the sun's centrality. For instance, in his letter to Herwart on March 28, 1605, the centrality of the sun was among several arguments to establish heliocentrism.[42] Thus although there are places where he did talk about the unique and divinelike position of the sun, so-called sun worship can hardly be seen as the main and sufficient reason for his adoption of heliocentrism.

A MORE COMPLETE EXPLANATION OF KEPLER'S ACCEPTANCE OF COPERNICANISM

Kepler's total, unconditional, and irrevocable adherence to Copernicanism required far more than "sun worship" or the argument from the comet. Additional material will have to be considered for a more complete and satisfactory explanation.[43] My investigations lead me to conclude that Kepler could make his brilliant move to accepting the hypothesis because he allowed rays of light to focus sharply on it from three different sources: science (empirical inquiry), philosophy, and religion. Empirical evidence alone could not have induced him to make such a move, for, as we will see shortly, Copernicanism had serious empirical problems. Similarly, philosophical ideas alone also could not have convinced him of the truth of this new hypothesis, for nothing novel is to be found in the philosophical arguments adduced to support it; such arguments had failed to convince many of his predecessors, and there is no good reason to expect a different result from the critical-minded Kepler. Nor could religious reasons alone have attracted him to heliocentrism. But in his thought and work he fostered an interlocking of these three fields of knowledge, which can in turn unlock the puzzle before us.

Scientific or Empirical Ideas

Kepler believed that a true, acceptable, scientific astronomical theory must be in agreement with past, present, and future observations. It should not only account for past and present observations of astronomers, but also predict accurately the future position and behavior of astronomical bodies; observational agreement was a crucial characteristic of any scientific theory. Kepler claimed to have found the Copernican theory meeting this requirement. He wrote right in the beginning of MC: "My confidence [in Copernicanism] was first established by the magnificent agreement of everything that is observed in the heavens with Copernicus's theory; since he not only derived the past motions which have been recapitulated from the remotest antiquity, but also predicted future motions, not indeed with great certainty, but far more certainly than Ptolemy, Alfonso, and the rest."[44] This statement, of course, was an exaggeration, because there were several observational data for which Copernicus could not account, for instance, the problem of the apparent size (brightness) of Venus. Furthermore, his claim that the new theory could yield more accurate results and better predictions was also an exaggeration, as in fact he realized (he recanted his claim). Later in the preface to his *Rudolphine tables* he admitted that the *Alphonsine tables* were as good as Reinhold's *Prutenic tables* based on Copernicus.[45] In any case, the fact that the new theory could account for phenomena as well as any other theory was a strong point in its favor. For if he could find other supporting reasons that were not shared by its rivals, then he could easily have rested certain in the new theory's superiority.

Maestlin's work on the comet of 1577[46] provided another empirical proof for Kepler.[47] This point is clear from Kepler's own words in this context: "Now on careful consideration of how easily the false disagrees with itself, and on the other hand how reliably truth is consistent with truth, a very strong argument for the Copernican arrangement of the spheres will quite rightly be drawn from that fact alone."[48] This passage is reminiscent of Maestlin's own reflection on his findings. His investigations on the comet revealed that "the comet chose no other place than the sphere of Venus itself."[49] Then Kepler's teacher asked rhetorically: "Could this calculation correspond in all its parts and with such perfection, if underlying it were false hypotheses?"[50] Indeed, he argued, since truth must be consistent with truth, a false hypothesis would not give true results and hence the Copernican hypothesis must be true. Quite obviously Kepler echoed the very same line of thought in MC. The argument here is flawed because, logically speaking, it is not impossible to get a true conclusion from false premises. Kepler was fully aware of this problem and wrestled with it at length in both MC and the *Apologia*. His basic solution was that a false hypothesis could yield a correct conclusion only accidently; its falsehood would betray itself when it had to confront other connected aspects of the problem.

Empirical evidence was one of the threee reasons he enumerated in his letter to Maestlin in 1595. "The second [reason] is physical, where I promise that I will defend the whole corpus of the Copernican hypothesis by means of principles of nature as being far more right than its opposite."[51]

According to Kepler, the most distinctive characteristic of a true scientific theory was its explanatory power, in the sense that such a theory could eliminate the "cause of wonder," the mystery, from phenomena by identifying the causes and showing how they followed from simple, intelligible causes in a consistent and coherent manner. Kepler believed that the new theory could do just that, whereas the old theory failed. This point of superiority impressed him more than anything else. He continued in MC: "However, what is far more important is that, for the things at which from others we learn to wonder, only Copernicus magnificently gives the explanation, and removes the cause of wonder, which is not knowing the causes."[52]

More specifically, what Kepler had in mind were a number of items noted in the ancient system which had remained uninterrelated and independent of each other. Astronomers had noticed that these cases depended on the sun in a significant way, but this reliance remained unexplained in a coherent manner. First of all, the size of the epicycles, according to the Ptolemaic system, showed a definite dependence on the sun: the size decreased with an increase of the distance from the sun. Secondly, it was found that the superior planets were always in opposition to the sun when they were at the bottom of the epicycles as determined by Ptolemy, and in conjunction when they were at the top. This meant that they were closest and hence brightest at opposition and farthest from the earth and hence dimmest at conjunction. Thirdly, as specified by the Ptolemaic system, the center of the epicycle of an inferior planet always lay on the earth–sun radius vector. On the other hand, in the case of the superior planets, the Ptolemaic epicyclic radius vector and solar radius vector always remained parallel to each other. All these points and other similar ones were stated as postulates by the Ptolemaic astronomers, who could not, however, explain why they were so and did not find any need for such an explanation.

Kepler found that all these cases could be accounted for as a matter of course in the new system on the basis of a moving earth and stationary sun. In MC he showed how this explanation could be carried out. The Copernican system could account for the unexplained puzzle in the Ptolemaic system, namely, the decrease of the size of the epicycle with an increase of distance from the sun. In the case of the inferior planets the ancients thought that the circular paths described by the planets around the central body were epicycles.[53] In the new system, since the circle of Mercury is smaller than that of Venus, when looked at from the position of the earth it looked smaller. For the superior planets the size of the epicycles decreased with distance because the angle subtended by the earth's orbit at them decreased with distance (see figure 1). Since these epicycles came about because of the motion of the earth, the angle produced due to the motion of the earth from A to B decreased with distance and hence the resulting observed displacement of the planets, CD and EF, also decreased with distance.

Kepler had no difficulty using the new model to explain what was observed at opposition and conjunction. The observation that at conjunction the planet Mars was always at the top of its epicycle can be accounted for: the variation takes place because of the motion of the earth (see figure 2). Let the earth move from A to B, giving rise to a conjunction when it is at B. At this point Mars will be on the opposite side of the sun from the earth, hence farthest from the earth and, therefore, dimmest. Similarly the situation at opposition can be explained.

That the centers of the epicyclic radius vectors always lay on the earth–sun radius vector followed from the Copernican idea that the inferior planets, like all the others, moved around the central sun. The earth, Venus, and Mercury moved in concentric circles.

We have already seen that despite his allegiance to Plato in several respects, Kepler remained a realist. From a realistic point of view, explanatory fertility, i.e., ability of a theory to explain not only what is originally intended but also other related points, is another significant characteristic of a true physical theory. According to this criterion, too, Copernicus scored very high; he could not only account for planetary motion but also reveal the order and relative distances of all the planets and predict a relative minimum radius of the sphere of the fixed stars.

FIGURE 1

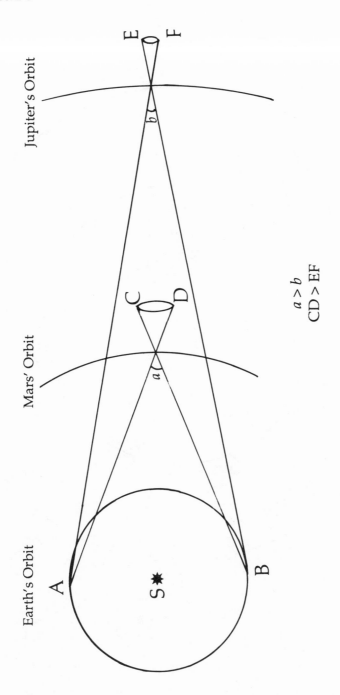

Figure 2

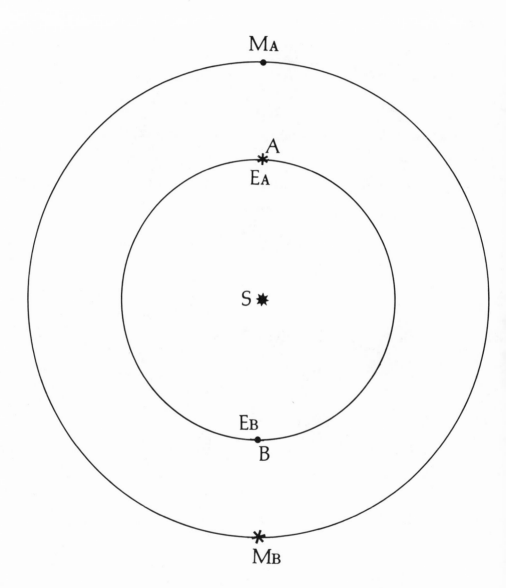

The above points, if true, could certainly argue for the strength of the new theory and may even show its superiority over all its rivals. The scientific advantages alone, however, could not have made Kepler a Copernican; several serious empirical problems were associated with the theory. I will go further: even if there were no serious observational objections at that time, the theory still could not have been acceptable on an empirical basis alone, because the positive evidence was not sufficient to induce one to choose it in preference to the old theories.

First of all, there were the old Aristotelian objections. Simple observations did not show a moving earth, but, rather, they attested to a moving sun. How could such a huge, heavy earth move? What would provide the immense force needed to do so? Contact force was necessary to move any material body. No one saw any such force acting on the earth. If the earth had an orbital motion, a stone dropped from a tower should not fall right at the foot, but rather a little behind. If the earth had an angular or rotational motion, objects on the earth must be thrown off.

Of course, the question can be raised as to whether these objections were really telling. The impetus theorists like Oresme had pointed out that the objections were not straightforwardly empirical. They could count against motion of the earth only if one assumed, as did implicitly Aristotle and his followers, including Tycho, that nothing would move unless acted upon by a contact force. Thus when a stone or a bird was not attached to the earth, no force would be acting on it, and so it would be left behind. Let us take these ideas for the moment as "empirical" (we shall analyze shortly why they are to be taken as such). Oresme argued that if the motion of the earth be somehow communicated to the object so that influence would remain in the object even when it was detached from the earth, it would be carried along with the earth and then the appearance would be just as we ordinarily observe. Hence, if the impetus theory were accepted, the objections would not have been telling. Nevertheless, the objections remained very strong in the eyes of many. Why? Because Aristotelian physics remained so powerfully convincing in comparison with the artificiality and sketchiness of impetus physics. Copernicus offered his own alternative physics (book 1), but except for Galileo it found no strong adherent. Kepler also would try to furnish a physics.

In the absence of an alternative convincing enough to sixteenth- and seventeenth-century minds, however, the Aristotelian objections seemed not to be arguments based on abstract and questionable principles, but true empirical objections.

Therefore the problem continued to challenge Copernicus. Tycho, Fabricius, Fludd, and others continued to bring these arguments in against the new view. For instance, Tycho proposed a hypothetical cannon experiment, which was nothing but a rerun of the old Aristotelian argument. According to him, if the earth moved, two cannon balls projected with the same force, under the same conditions, but one to the east and the other to the west, should not cover the same length of space on the surface of the earth; observations showed that they did. To answer these and similar problems a new physics had to be developed. In the absence of such a satisfactory explanation of these difficulties, adopting the Copernican system meant "committing oneself to an hypothesis that appeared to be patently refuted by a well-supported theory of terrestrial physics."[54]

With regard to accuracy of prediction also the new theory seemed to have fared poorly. It could not predict more accurately in the case of every planet and in every aspect of the planet than the Ptolemaic model could, especially the then-updated version of Peurbach. Kuhn points out that the accuracy of the new theory compared with the simplified twelve-circle version of Ptolemy.[55] Gingerich also arrives at the same conclusion and argues that reason and insight had to come in to effect the Copernican revolution: "Copernicus had tried to effect a cosmological revolution without really raising the standards of prediction; his was a vision of the mind's eye, not the result of an observer with his quadrant and armillary sphere."[56] Although fourteen centuries separated Copernicus from Ptolemy, the accumulated experience did not enable Copernicus to come up with a substantial advance in the predictive mechanism.[57]

Pretelescopic astronomers, confusing apparent brightness with large size, created another nagging problem, especially serious in the case of Venus. Because of its great brightness it seemed (incorrectly) to the naked eye large in size. Thus the observed variation in the apparent size of Venus posed a problem to the new theory. When in superior conjunction, Venus was slightly more than six times distant from the earth than when it was at inferior conjunction. Hence its apparent size (i.e., apparent surface area) should be nearly forty times as great at inferior conjunction. However, it was found that observation by naked eye revealed not even a doubling of its apparent size. This was a strong objection to Copernicus. To be sure, with the advent of Galileo's telescopic observations this objection was removed, but until they were recognized by the scientific community, it posed a serious threat to the Canon's innovations. Our problem, in any case, remains, since Kepler accepted heliocentrism before such telescopic observations were available. These problems prompted Hanson to argue that the genius of Ptolemy should certainly have considered the possibility of a moving earth. Its simplicity must have appealed to him. Since he was concerned about "saving the phenomena" only, all that he looked for was a better computational mechanism. He could with impunity have entertained a system with a moving earth; he did not, because observational facts, as we have explained above, would not permit him to do so. "Observational facts, not divine principles or fear of heresy, prevented Ptolemy from becoming a heliocentrist."[58]

Copernicanism not only failed to account for certain observational results, but it also gave rise to new problems that at that time the theory was helpless to handle. If the new theory was correct, then we must observe stellar parallax, i.e., the apparent position of the fixed stars must shift when observations are taken from two different positions of the earth's orbit separated by a long distance. No such parallax was found; not even Tycho, the observational genius of the day, could find any with the best of his instruments. This meant that either the earth did not move or that the fixed stars were too far away from the earth to give rise to any observable shift. Copernicus chose the second explanation, but not even his best supporters like Kepler were satisfied with such an explanation.

Tycho considered the nonobservability of the parallax an extremely serious objection, as he made very clear in his correspondence with the German astronomer Rothmann.[59] According to the Dane's calculations, in Copernicus's theory the Saturn–fixed star distance would be seven hundred times the Saturn–sun distance. He considered such a result to be incredible. He raised another serious objection, too:[60] if Copernicanism was accepted, then a star of the third magnitude, the apparent size of which was assumed to be one minute, would be as large as the orbit of the earth. This also seemed unbelievable for Tycho.

Philosophical Ideas

The discussion above shows that empirical considerations alone could not have motivated any critical-minded person in the second half of the sixteenth century to embrace the new theory. Kepler needed more than that to be induced to accept Copernicanism. We can show that he had several philosophical reasons in favor of the new theory.

Kepler found that the new theory "did not offend against nature,"[61] meaning that the theory was in agreement with natural philosophical principles such as "nature loves simplicity,"[62] "nature loves harmony," "nature is economic," "nature loves unity."[63] He explicitly discussed how Copernicanism was consonant with these principles. For instance, the new theory satisfied the principles of simplicity in both its meanings, systematic and structural.

Systematic simplicity was characterized by a system which could account for the maximum number of phenomena with a minimum of postulates. Eleven different motions had been involved in the old system: the three motions that had been introduced by the ancients to account for the observed motions of Mercury; the three large epicyclic motions of Saturn, Jupiter, and Mars; the five different motions attributed to the five planets by Ptolemy to account for their variation of latitudes. Where the old system needed all eleven motions, however, the new heliocentrism could do with just one. Heliocentrism was therefore simple because by this theory "all these motions, eleven in number, are banished from the universe by the substitution of a single motion of the earth. . . ."[64]

The new theory also showed, in a way, structural simplicity which consisted essentially in getting rid of complicated epicyclic arrangements. Copernicus brought this point out when he wrote: "I hold it easier to conceive this [the heliocentric view] than to let the mind be distracted by endless multitude of circles, which those are obliged to entertain who detain the earth in the center of the world."[65] Kepler seemed to have agreed with this claim; according to him, Copernicus "freed nature from the burdensome and useless paraphernalia of all these immense circles. . . ."[66] Nevertheless, it is incorrect to say that Copernicus got rid of all the epicycles; he could eliminate only the large ones in the Ptolemaic system having to do with retrograde motion.

Another area where the old system had become complex was in the triplication of the central reference points: a center around which the planet's angular velocity was constant (i.e., the equant center), a center around which epicyclic distance was constant, and a center around which planetary observations were made. The new system asserted that it could get rid of such a triplication. But how justified was this and other such claims?

Kepler was convinced that nature is rational and so everything in it should have a sufficient reason. Any theory that claims to deal with the structure of the universe must provide reasons for that particular structure. According to him, in some ways the new theory met this criterion of rationality, because it could give "reasons for the number, extent, and time of retrogressions, and why they agree precisely, as they do."[67]

Kepler believed that nature is a unified whole: everything in nature is related to every other phenomenon so that no part can be altered or removed without affecting the rest. A true theory of nature must reflect its harmonious interconnectivity. Copernicus claimed that his system possessed such a harmonious unity and marvelous symmetry, a claim with good justification. For instance, in the new system the relationship between the distance of each planet from the sun and its period of revolution exhibited a regular pattern; the period of revolution increased with distance; no two planets had the same period. The ordering of the planets was clearly specified. On the other hand, in the Ptolemaic system, the superior planets followed a pattern, but Mercury, Venus, and the sun all required only one year to complete their journey around the earth; the order of the orbits had therefore been a matter of debate. In the new system the orbits of the earth, Venus, and Mercury could be uniquely determined—the time of revolution of Venus calculated to 225 days, and of Mercury, 88 days—hence Mercury is closest to the sun, followed by Venus and the earth. The new system was neat and followed a regular pattern. No planet's distance or motion could be altered without affecting the others. As Copernicus put it, "All these orbs and the heaven itself are so connected that in no part can anything be transposed without confusion to the rest and to the whole universe."[68] There is no doubt that Kepler believed that the heliocentric theory had met the criterion of interconnectivity. As he remarked, in addition to getting rid of the immense epicycles "Copernicus has opened to us an inexhaustible treasury of calculations on the fitting together of the whole universe and of all the bodies in it."[69] On the other hand, in the Ptolemaic system each planet was treated independently of the others and one could be changed without affecting the rest. Naturally, the harmonious unity that Kepler found in the new theory must have had a telling impact on him.

Further, a true theory, Kepler argued, must withstand any kind of test, must be, as it were, "omni-probable." A theory should satisfy any demand put to it and give the true answer, irrespective of how the demands are put, how the tests have been designed. He held that the new theory had this versatility of response to demands.

> If you tell him [a Copernican] to derive from the hypothesis, once it has been stated, any of the phenomena which are actually observed in the heavens, to argue backwards, to argue forwards, to infer from one motion to another, and to perform anything whatever that the true state of affairs permits, he will have no difficulty with any point, if it is authentic, and even from the most intricate twistings of the argument he will return with complete consistency to the same assumptions.[70]

However, there were difficulties with these claimed virtues of the system since serious counterclaims could be raised. For instance, one could easily challenge the claim to simplicity. Scholars have pointed out that the updated system of Peurbach could manage with fewer epicycles than the Copernican theory. The large Ptolemaic epicycles were indeed absent in the new system, but it still retained smaller ones to account for the different aspects of planetary motions.[71]

Not only in the number of the epicycles but also in the explanation of the mechanism of planetary motion this theory was far from simple. For instance, consider Copernicus's account of the motion of the earth.[72] It was a far cry from the simple structure we would expect (see figure 3). The sun was fixed at S and the earth moved along the eccentric. The center of earth's eccentric O_E was a moving one. O_E revolved slowly about a point O, which in turn revolved on a sun-centered circle. All these arrangements, necessary to get a satisfactory system, were certainly far from the simple heliocentric system Copernicus seemed to have promised to deliver. The motion of the other planets, too, had complicated structures.

FIGURE 3

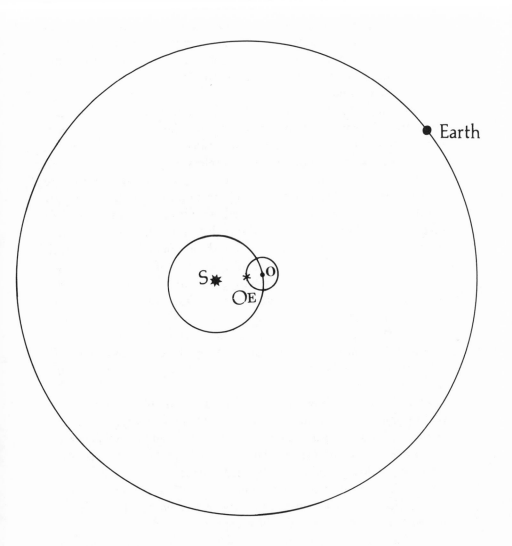

Another failure is also brought out by figure 3: the system was not in fact truly heliocentric, in the sense that the sun was not at the center of the universe. In fact, only in DR book 1 does the sun occupy the central position (even this point is arguable, as I shall discuss in the next chapter). In books 2–6 the sun is not at the center of the universe but, rather, at a distance from it. The central point is vacant. No triplication of the central reference points occurs, but they are duplicated. The claimed simplicity of the new system left much to be desired.

Copernicus used the principle that nature is economical to strengthen the superiority of his system. According to him, his theory took into account the fact that "the wisdom of nature is such that it produces nothing superfluous or useless. . . ."[73] But Tycho and others used exactly the same principle to expose the weakness of the new theory. The Dane argued that the nonobservability of stellar parallax meant that the Saturn–fixed star distance must be at least seven hundred times the Saturn–sun distance. This implied that a huge, wide expanse of space existed with no apparent purpose. Such a vast empty space went against all principles of economy and was an insult, even, to the infinitely wise God who created it. God could never commit such a colossal mistake. The new theory must be wrong headed.[74]

Furthermore, the theory could not meet Kepler's strongest criterion that a true scientific theory should meet all the demands of test. His claim that heliocentrism could satisfy this demand was only wishful thinking.

Our discussion shows that Kepler had available to him several philosophical reasons not only in favor of the new theory, but also against it. Philosophy alone could not have led him to a wholehearted acceptance of heliocentrism. This would have been so even if no opposing philosophical ideas obtained because other theories existed that could also boast of philosophical virtues. The Eudoxian model, with all its concentric spheres, must have seemed simple and harmonious. Yet Kepler did not find it acceptable.

We have seen that neither empirical nor philosophical concerns alone were sufficient for Kepler's choice. Could they together have made him a Copernican? When the two were put together, the arguments in favor of the theory undoubtedly became more powerful. The combined power of empirical and philosophical reasons could establish superiority over the Eudoxian system because the new theory was empirically superior. This combined power also could enable the new theory to establish its superiority over the Ptolemaic since the latter scored rather low from the philosophical point of view (e.g., from considerations of simplicity, unity, etc.). However, the combined power was not enough to make Kepler an irrevocable adherent of heliocentrism. For one thing, the system of Tycho enjoyed at that time most of the advantages attributed to Copernicanism. In fact, as we have seen already, the two theories were mathematically equivalent. Although the Tychonic system was a compromise between the old and the new, it had not sacrificed any of the major mathematical harmonies of the Copernican system. It also had considerable physical advantage because it did not require the earth to move. Furthermore, it remained faithful to the pertinent scriptural passages as they were then officially understood. Thus even if one were to put together the empirical and philosophical reasons in favor of the new theory, they were not strong enough to prove that it was the only true theory. There had to be further reasons in its support. I propose that religious ideas were also crucial for making Kepler a Copernican.

Religious Ideas

Kepler believed that the Copernican theory was unique and uniquely true. It and it alone furnished a satisfactory physical explanation of the observed and observable natural phenomena. Since at that time neither empirical nor philosophical arguments taken singly or in combination could establish this claim, he had to relate Copernicanism to something that he considered unique. Because for him God was a uniquely true being, if the theory could be related to God in a special and natural (nonartificial) way, then it must be true. In fact, Kepler himself expressed this point in the introduction to MC: "That I dared so much was due to the splendid harmony of those things which are at rest, the sun, the fixed stars, and the intermediate space, with God the Father, and the Son, and the Holy Spirit."[75] He found a close relationship between the Copernican world structure and the Holy Trinity.

In part 1, we have discussed Kepler's belief that the universe is the image of God; for Kepler the material universe is a reflection of the divine. He found that the Copernican model could support these beliefs as no other model could. The Copernican structure of the world with the sun at the center, the fixed stars at the extreme surface, and the planets in the intervening space was the true representation of the Trinity. He saw a one-to-one correspondence between the different parts of the Copernican universe and the persons of the Trinity: the Father with the sun at the center, the Son with the surface, and the Holy Spirit with the intervening, dynamic, emanation-filled space. Since the Trinity is something unique, this relationship rendered the Copernican system something unique. Furthermore, the unique place given to the sun in this theory corresponded to the unique place God the Father occupies in the heavenly world. In this way the new theory supported Kepler's religious beliefs in a remarkable way.

The similarity between God the Father in the Trinity and the sun in the material world was not confined to their paramount *position* in their respective "structures" but extended also to their respective *operations* and *functions*. Kepler argued that just as creation is the characteristic function of God the Father, so is giving motion the special function of the sun. He made this point explicitly clear in a letter to Maestlin in 1595, where Kepler drew a beautiful parallel between the sun, the unmoved mover, and God the Father, the uncreated Creator. In the solar motive force that moved the planets he saw a reflection of the creative power and activity of God the Father: "The sun disperses a moving power through the medium in which are the movables, and in just this way the Father creates through the Spirit or through the power of the Spirit."[76] When a theory could sustain such a striking parallel with a truth he firmly believed in, it should be no wonder that he took such a theory as true.

That the comparison between God the Father and the sun played a decisive role in Kepler's acceptance of heliocentrism is clear from other statements as well. He enumerated in his letter to Herwart six reasons for adopting the new theory. The second was based precisely on this comparison between God the Father and the sun. "The center is the origin and beginning of the sphere. Indeed, the origin has precedence everywhere and is by nature always the first. When we apply this consideration to the most Holy Trinity, the center refers to the image of God the Father. Hence the center of this material world-sphere should be adorned by the most ornate body, that is the sun, on account of light and life. . . ."[77] This passage made clear that just as God the Father is the source of everything and at the center of everything, the sun also must be the source and center of the material world. Just as the center of the Trinity is occupied by God the Father, the highest majesty and supreme beauty, so also the center of the sphere of the material universe should be adorned by the most ornate body. Kepler held that the sun alone could take that place because it alone was the source of life and light. Being the source of light, it was the most beautiful body in the universe and hence most fit to be the representation of God the Father. Furthermore, as we have seen already, in Christian theology God and the source of light and life have been very closely and intimately related. Hence the sun which Kepler identified with the source of light and life was most suitable to be the counterpart of the invisible God the Father in the visible universe. It follows, therefore, that a true theory of the universe must give the sun such a unique position. Since only in the Copernican theory, and not in the Tychonic system, did the sun occupy such a position, Copernicanism alone must be true. This same religious idea automatically ruled out the Tychonian system, the prime rival to Copernicanism.

The importance of the sun was decisive in Kepler's adoption of the new theory but could not be considered the most potent or the main and sufficient reason, for it was only one of the six reasons he gave to Herwart. Kepler never said that this was his main and sufficient reason. Religious reasons alone could not have motivated him to embrace heliocentrism.

I have argued that philosophical and religious reasons were not just a "rescue squad" but, rather, "respectable partners" in Kepler's thought and works. His philosophical views did not begin to function only when the empirical failed. Nor did religious reasons operate only when the other two factors were spent. To arrive at Kepler's conclusion all three domains had to collaborate.

Kepler's basic position can be stated as follows: Copernicanism was the only true theory of nature that could account for observed and observable phenomena in accordance with accepted philosophical principles. To reach such a conclusion, arguments from all the three sources—empiricism, philosophy, and religion—were necessary. The empirical arguments could show that the theory could account for the observed and observable phenomena. Observational data could not establish alone, however, that Copernicanism was the *only true theory* since other systems, like the Ptolemaic theory, especially as modified by Peurbach, could also account for such phenomena. But when philosophical ideas were introduced, such systems could indeed be ruled out because they could not adequately satisfy the principles of harmony, unity, economy, etc. Nevertheless, even when the empirical and philosophical consideratons were put together, the Tychonic system still could not be dismissed; in fact, to many scholars it was the better contender. It lacked a certain amount of symmetry because of its "duplication of centers of motion," i.e., the five planets around the sun and the sun and moon around the earth. Then again, Copernicanism had its own problems. Hence it was not possible to place one theory over the other and, much less, to accept one as uniquely true. To make such a move religious ideas had to be considered. Only Copernicanism could satisfy Kepler's belief that the heavenly, divine world mirrored the material one. Only the Copernican system placed the sun, the representation of God the Father, at the center. Only this system best expressed the comparison between the Holy Trinity and the spherical universe. Kepler believed that since God is supremely true, a system that reflected the Godhead must also be true. Thus his religious ideas could show that the Copernican system was true and uniquely true.

In this discussion I considered empirical factors first, but I could have begun with any one of the three areas that influenced Kepler's acceptance of Copernicanism, since the specific contributions of each sphere were unique to each, not mutually exclusive.[78] In this step of Kepler's process of discovery, the complementarity of empiricism, philosophy, and religion was very clear. As we shall see, in some other steps the complementary roles would not appear so sharply, but they were certainly alive and active.

5

THE DEVELOPMENT OF A TRULY
HELIOCENTRIC VIEW

Although the shift from the geocentric to heliocentric worldview is almost universally attributed to Copernicus, his system was not in fact characterized by the sun being exactly at the center. The sun certainly was not the dynamic center of the universe, nor even the geometrical center (the dynamic center refers to a body's function or activity, whereas the geometrical center refers to the position or location of a body). It is highly doubtful whether the sun occupies the center even in book 1 of DR because Copernicus betrayed clear signs of wavering, as in chapter 10: "Circa ipsum (solem) esse centrum mundi," which Rosen translates as "Near the sun is the center of the universe."[1] The use of *circa* clearly shows that the sun does not occupy the central position of the universe. Later in the same chapter, however, Copernicus wrote: "In medio vero omnia residet sol" ("in the middle of everything is the sun").[2] Even in book 1 Copernicus was not very consistent, although some scholars[3] still argue that in book 1 the sun is taken as the center of the universe (as we shall see, these scholars simply are not quite accurate).

The Copernican system was not truly heliocentric even from a geometrical point of view. Rather, the sun, like a lamp illuminating the surrounding space, illuminated the universe but did not govern it, despite Copernicus's claim that the sun was the supreme king seated on his royal throne. Even if we discount these problems and subscribe to the view that in the Copernican system the sun was the geometrical center, that position left the sun a far cry from being the true dynamic center of the universe, far from the real center to which everything converged and from which everything diverged.

Kepler, on the other hand, could never be satisfied with a world-view which failed to consider the sun beyond its being a geometrical center. He placed the sun really at the center of the universe. Clearly he was the first to give a truly heliocentric worldview.

The creation of the truly heliocentric view involved a number of steps. Acceptance of Copernicanism was only a first necessary condition; heliocentrism needed much more. Since it demanded making the sun both the geometrical and dynamic center, heliocentrism required the following additional steps: (1) the shifting of the point of reference to the real body of the sun, (2) the acceptance of the sun as the point of intersection of planetary orbital planes, (3) the establishment of the fact that the orbital planes have a constant inclination to the ecliptic, (4) the reintroduction of the equant, (5) making the sun the true dynamic center of the system. In this chapter I shall discuss the first four points together because they are intimately interrelated; they all refer to the position of the sun as the true geometrical center. The fifth point, on the other hand, refers to the function of the sun as the dynamic center, so I shall take up this issue separately, in chapter 8.

SHIFTING THE POINT OF REFERENCE TO THE REAL SUN

The Discovery and Its Importance

Kepler shifted the point of reference to the real sun, to the center of the body of the sun. He wrote to Maestlin: "I hold as most certain that the eccentric is arranged about the true body of the sun and not about the central point of the Copernican solar [orbit]."[4] This idea was no minor matter, since it went against the practice of great authorities before him: Ptolemy took the center of the earth as the point of reference, whereas Copernicus and Tycho took the mean sun. As Kepler himself remarked in AN, "Copernicus and Tycho followed Ptolemy, carrying over his assumptions. I, as you see in chapter 15 of my *Mysterium Cosmographicum*, take instead the apparent position, the true body of the sun, as my reference point, and will vindicate that postion with proofs in parts four and five of this work."[5] Again, Copernicus argued that this point was not fixed, but librating. Gingerich, commenting on Kepler's shifting the point of reference, remarks: "The idea is so important that we should perhaps call it Kepler's zeroth law. It is a sign of Kepler's genius that he so quickly recognized this as the crucial first move in the reform of astronomy."[6] Small brings out the innovation's importance: "In fact, excepting only the system of Copernicus, it was an improvement more important, and of greater consequence, to simplify the science, than any which had been introduced in all the preceding ages; and his successful and decisive establishment of its truth and propriety, may be justly ranked among his greatest discoveries, and equally deserves our attention with those which have been more generally celebrated."[7] The move was indeed a significant contribution by Kepler, although it is debatable whether this simplification must be counted among his greatest discoveries. It gave the correct and necessary starting point; and insofar as getting the right start is crucial for any scientific work, this was a decisive step on his way to discovering the laws of astronomy and gave him direction and guidance as to how to proceed in his investigation. From then on, the sun's position and properties began to play a central role in his astronomy.

Kepler's move had significant consequences. It necessitated the displacement of the line of apsides (these points can be illustrated by figure 4). Let O be the center of the earth's orbit and let the eye be placed at O. According to the traditional position, O will be the point of reference. Let observations of the planet be made in the points J, H, K of the planetary orbit. The corresponding points on the earth's orbit are G, A, C. The sun is at point S. According to Ptolemy, E will be the equant and OER will be the line of apsides. On the other hand, in Kepler's theory the line of apsides will be SEL and the eccentricity SE. The new move, therefore, involved a displacement of the line of apsides; it also destroyed the uniformity of rotation of the planetary spheres which Copernicus so triumphantly claimed to have achieved.

FIGURE 4

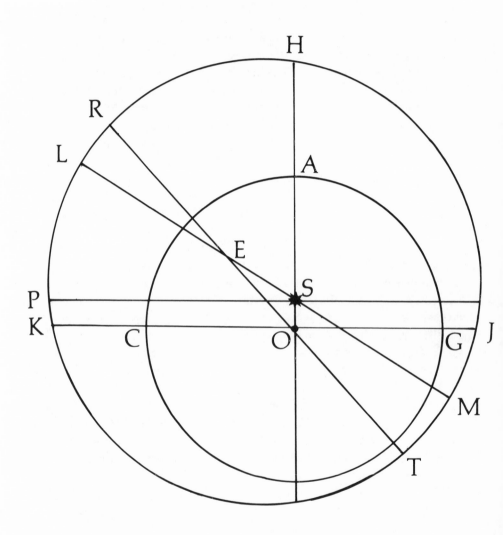

Scientific or Empirical Ideas

When Kepler made the initial discovery that the real sun should be the point of reference, he lacked firm empirical proof. However, even at this stage he had some empirical evidence. In the MC he had referred the eccentricities to the real sun rather than to the mean sun.[8] Already in 1600 he reported that his study of Mars using Tycho's observations confirmed admirably MC in two positions. The agreement with Tycho's observation assured Kepler that he was on the right track. This point he conveyed in a letter to Herwart: "Since in it [MC] I had referred the eccentricities of all the planets to the body of the sun, I greatly feared that Tycho, like Copernicus, would refer it to the mid-point of the sun's (earth's) orbit. But Mars constantly rejected all points other than the center point of the sun."[9] Later in 1604 while writing to Fabricius, Kepler gave another reason based on empirical considerations to show that all distances must be referred to the real sun. In his report on the study of Mars on the basis of the distances of the planet he wrote: "From the line of apogee of the sun nothing remains in excess in the eccentric of Mars if you refer it to the sun itself. But if you refer it to the point or position of the mean sun, something does indeed remain in excess. It is not large as far as the locations are concerned, but it is greater for distances. And I have included this among the reasons why I refer this theory to the true center of the sun."[10]

Although the above considerations strengthened him in his belief, the decisive empirical proof he presented much later, in AN, chapter 52, the very title of which conveyed an important message to his readers: "A demonstration by means of the observations of chapter 51 to show that the eccentric of the planet [Mars] is not arranged around the center of the sun's epicycle or around the point of the mean position of the sun, but around the very body of the sun. . . ." In chapter 51 of AN Kepler had established that the distances from the sun to the planet at corresponding positions in the two opposite semicircles (or semiovals) of the orbit were equal. In chapter 52 he showed through a geometrical argument that this equality demanded that the distances be referred to the real sun, and not to the center of the earth's orbit. Kepler concluded the chapter as follows: "Once more I have kept my pledge made in chapter 6 and throughout this work without in any way begging the question, and I have taught that the eccentric of Mars cannot be referred to anything except the sun itself. Therefore not only reason, but also observations stand by me as long as I refer observations of Mars to the apparent motion of the sun, rather than to its mean motion."[11]

However, all these empirical data came long after Kepler had arrived at the idea. As we have seen, he had insisted on making the true sun the point of reference already during the days of MC. Hence empirical considerations could not have played the only decisive role in the genesis of this idea. Furthermore, his initial empirical application of the idea gave him very discouraging results. In the MC, when he found that the Copernican data did not fit satisfactorily with the polyhedral theory, he argued that the root of the problem was with the Copernican data which were based on the assumption of the mean sun rather than the real. Hence he asked Maestlin to recalculate the data by referring these distances to the true sun; this his teacher promptly did. However, to Kepler's great surprise and disappointment, he found that the new values gave hardly any better results. Yet he refused to give up the idea. Hence in the case of the origin of this important innovation, nonempirical factors rather than empirical ones must have been crucial.

Philosophical Ideas

Several philosophical considerations prompted Kepler to depart from the past to introduce the innovation of shifting the point of reference to the real sun. The old view of Copernicus and Tycho violated the principles of realism, causality, simplicity, and economy. There all distances and motions were referred to an empty point, so planetary phenomena were dependent on a point that was just nothing, which obviously went against the principle of realism. Moreover, Kepler believed that the different items in the universe were causally connected; an empty point could not be a cause since "nothing" cannot be a cause of something, so the Copernican position broke the principle of causality. Further, according to Copernicus, the point of reference was not a fixed, but a librating one; even if it were possible for the intelligences to perform the motion with regularity, extremely complicated calculations would have to be done, which put added demands on the movers: such a multiplication of functions and complication of motions violated the principles of simplicity and economy.

On the other hand, if the center was a physical body, if that was fixed and of great magnitude, then the problems above could be solved and planetary motion could be causally explained. As Kepler put it, the sun "which has a geometric affinity for motion and is invested with a body of no small magnitude"[12] could be a satisfactory solution, whereas a point that was not fixed which "sometimes precedes, sometimes follows the sun, at one time above the sun, at another below it . . . "[13] only multiplied the problems. According to him, the stationary sun could provide an adequate physical cause that could explain the relevant planetary phenomena.

At the same time, a few philosophical considerations seemed to counter Kepler's innovation. For instance, his view entailed abandoning the idea of planetary motion around the central point of the circular orbit, because in the new system the heavenly bodies went around an off-center body. The innovation seemed to have destroyed the symmetric, harmonic arrangement that was considered to be one of the finest virtues of the Copernican system. For Kepler, too, symmetry was important, since, assuredly, it must have been one of the strongest reasons for his stiff resistance to giving up the circular orbit.

Religious Ideas

The discussion in the previous section shows that empirical and philosophical considerations were not fully sufficient for Kepler to go against the age-old traditions of many authorities because there were objections against his innovation. Further support needed to take such a momentous step came from religious ideas.

According to Kepler, the sun and the sun alone must be the body serving as the point of reference. The sun was not just one among several choices, it was the only choice and nothing else could have taken its place. How did he know that the sun had such a unique place in the universe? How could he know the sun could perform such a unique function? At this time neither empirical nor philosophical considerations could have established the sun's uniqueness. But he could accept the sun as solely fit for the purpose because he saw in the sun the reflection of God the Father; just as God the Father's place in the Trinity is unique, so also was the sun's position unique in the solar system. And as God the Father is—functions as—the origin of everything, the source of all energy, the source of life, so too, similarly, the sun was the source of life and light.[14] These ideas gave Kepler the confidence to believe that the sun was perfectly and solely qualified to represent God the Father and hence to be the origin to which all aspects of planetary motion could be referred.

That Kepler believed that the sun, like God the Father, must be the center was evident from a teleological argument. In Kepler's view, the very nature and function of the sun demanded that it occupy the central position. He wrote in *Ad Vitellionem paralipomena*: "Hence the sun is a certain body in which [resides] that faculty of communicating itself to all things which we call light. For this reason alone its rightful place is the middle point and center of the whole world, so that it may diffuse itself perpetually and uniformly throughout the universe."[15] Therefore the sun could not possibly occupy any place other than the center of the universe. Just as the Father by very nature and function has to be the center of the heavenly world, the sun by nature and function had to be the center of the material world.

ACCEPTANCE OF THE SUN AS THE POINT OF INTERSECTION OF PLANETARY ORBITAL PLANES

The Discovery and Its Importance

Kepler believed that the orbital planes of the planets should pass through the body of the sun. Acording to him, the planes of "all the eccentrics intersect in no other place than the very center of the solar body (not some nearby point), contrary to what Copernicus and Brahe thought."[16] Kepler's move here also broke with the past because it differed decisively from his predecessors' approaches. In the Ptolemaic system the planetary planes were made to intersect at the center of the earth, whereas in the Copernican system they met at the center of the earth's orbit. Since Kepler's new idea meant that the body of the sun was the meeting point of all planetary planes, the innovation was also an important step in the creation of a truly heliocentric system.

Scientific or Empirical Ideas

As in shifting the point of reference to the real sun, empirical evidence for accepting the sun as the point of intersection of planetary orbital planes also came quite late in Kepler's work on Mars. Chapter 6 of AN offered some preliminary scientific findings; in his view the considerations here "opened the door" for further empirical evidence.[17] At first he argued that if the planetary planes were posited to pass through the sun, then the inclination of these planes to the ecliptic would remain invariant, and in AN he established this invariance of the inclination of the planetary planes. These studies could provide some support to his innovation. The most direct and decisive proof came later, however, in chapter 52 of AN.[18] Here he proved that except for those points lying in the line of apsides passing through the center of the sun, there was none within the orbit of the planet from which equal straight lines could be drawn to any two such corresponding points. Hence the plane of the planet should pass through the center of the sun.

Philosophical Ideas

In AN Kepler left no doubt that he had arrived at this idea of the sun as the point of intersection long before he could get proper empirical data; he himself admitted that he "had deduced it a priori."[19] His philosophical principles of realism and unity must have enabled him to make the deduction. Realism demanded that the orbital planes pass through a real body, rather than through a mere mathematical point, as in the case of the Copernican system. This innovation agreed with the principle of unity, too, for all the planetary planes were related to one body, i.e., the sun, and all were supposed to meet at one single body. Thus the discovery was helpful in building up a unified planetary system.

Religious Ideas

One of the reasons why Kepler could posit his new idea in the absence of proper empirical evidence was that it helped him to make the sun the true center of the universe. This innovation obviously was very much in keeping with his fundamental belief that the sun was uniquely fit to be the visible representation of the invisible God the Father.

ESTABLISHMENT OF THE CONSTANCY OF INCLINATION OF PLANETARY ORBITAL PLANES TO THE ECLIPTIC

The Discovery and Its Importance

Copernicus had investigated the inclination of the orbit of planets to the ecliptic. However, as Kepler pointed out, Copernicus made use in his study of Ptolemy's data for his observations of latitude, and out of deference to the great astronomer never subjected the data to any critical investigation. These studies led Copernicus to conclude that the inclination of the orbits of the planets to the plane of the ecliptic was variable. Since it was accepted that the plane of the ecliptic was stable, he believed that planes of the orbits of the planets oscillated backwards and forwards, each in its line of nodes. According to him, the orbit of Mars librated through an angle of 1 degree and 41 minutes and this libration was such that the inclination always increased in proportion to the increase of anomaly. Kepler was critical of his master for his wrongheaded approach and remarked that Copernicus's ignorance of his own richness misled him into taking such a misguided step.

Right from the beginning Kepler believed that the inclination of the orbits to the ecliptic should be constant. By using three different methods to ensure that he was correct, he determined the inclination of the plane.[20] The first method, applicable to all planets but Mercury, was based on observations of latitude made in the limits (i.e., points farthest removed from the ecliptic) when the planet–earth and planet–sun distances were equal. It presupposed former theories for the determination of the mutual ratio of the orbits and the heliocentric distances of the planet. For Mars this gave a value equal to 1 degree and 50 minutes. The second method required special conditions but presupposed no former theory. This method required that the observation of the planet be made when both the sun and the earth were in the line of its nodes and the geocentric place of the planet was at 90 degrees from the nodes; obviously, under these conditions the observed latitude directly gave the inclination sought for. Tycho's abundant treasure of observations furnished four such examples, which approximately satisfied the requirements. This method also yielded the same value for Mars of 1 degree and 50 minutes. Finally, Kepler used the method employed by Copernicus himself, which also gave the same result. From these three methods Kepler concluded that for Mars the inclination of the plane of the eccentric to the ecliptic plane was stable and constant. Since he had observed the invariance of inclination not only for Mars, but also for Venus and Mercury, he unhesitatingly generalized the results to a law that the inclinations of the planes of all the planets to the plane of the ecliptic are invariable and constant.

Kepler considered this discovery as his first victory over Mars. According to Small, the finding "may justly be ranked among Kepler's great discoveries."[21] The result was undoubtedly another necessary step for his further research in planetary motion.

Scientific or Empirical Ideas

What factors brought about Kepler's first victory over Mars? The usual and, in some ways, the most obvious answer is to call attention to his empirical investigations. For he discusses the invariance of planetary planes to the ecliptic in chapters 13 through 15 of AN and the whole discussion emphasizes the role of observation. Although observational data played a crucial role in this discovery, a careful study of Kepler's words and deeds reveals the philosophical and religious considerations that were also involved.

For one thing, Kepler professed that he had the idea before he could study Tycho's data. For instance, in his letter to Magini in Bologna on June 1, 1601, Kepler vehemently criticized Copernicus and affirmed that the inclination of the orbit was stable and constant.[22] This letter seemed to indicate that he had the idea already in MC, or at least soon afterwards. That the idea predated his study of Tycho's data is also clear from his statement in AN that he was suspicious of the old view and "always fought against it, although as yet I had not seen Tycho's observations."[23]

Observation and empirical considerations seem to have functioned here as firmly establishing his hunches and intuitions. Thus immediately after mentioning his persistent objection to a librating orbit, he said: "Wherefore I congratulate myself all the more because observations have been found to support me, as in the case of many other ideas conceived by me prior to observations."[24] He congratulated himself because once again he had found that his hunches had been vindicated. According to him, as had happened many times in the past, observations stood by his "preconceptions"; i.e., ideas conceived before he had adequate observational evidence were upheld by subsequent scientific research. Hence nonobservational factors, such as philosophical principles and religious beliefs, must also have contributed to the origin of his idea.

Philosophical Ideas

A number of philosophical principles seemed to have been active in the initial discovery of the constancy of inclination. Obviously, a librating orbit violated the principles of simplicity and economy: simplicity, because such a system was unnecessarily complex; economy, because the arrangement involved the expending of unnecessary work. Kepler's own statement implied that these principles were crucial to his initial discovery. He characterized the Copernican position as a meaningless complication of the various orbits,[25] as totally unnecessary and useless: he even referred to his master's position as monstrous.[26] Kepler's statements point to the old view's flagrant disregard for the principles of simplicity and economy.

But could concern for simplicity and economy alone have given him enough confidence to go against tradition, against his own master, against observational data considered to be reliable at the time? He needed more, and the additional argument was supplied by the fact that establishing constancy of inclination would strengthen the centrality of the sun.

Religious Ideas

Since Kepler considered the ecliptic the plane of the sun,[27] constancy of inclination to the ecliptic meant constancy of inclination with respect to the sun. With the acceptance of this point it followed that the sun was the body to which were referred not only all distances, but also all inclinations of the planetary orbits. Constancy of inclination meant that the sun was the point of reference not only of all linear distances, but also of all angular distances. We know that in mathematics the position of a body can be defined fully in terms of the distance from the origin and the angle it makes with the origin (cf. polar coordinates). By arguing that the distances and angles of the planet must be referred to the sun Kepler could ensure that the position of the planets was defined or specified in relation to the sun. Hence this move once more highlighted the preeminent position of the sun in his system of thought. As we have already noticed, a theory that elevated the significance of the sun had a special attraction for him because the sun was the visible representation of the invisible God the Father. Hence this new idea preserved the principles of simplicity and economy and affirmed the supremacy of God in the created universe. This discovery, like the others, also involved religious ideas.

REINTRODUCTION OF THE EQUANT

The Discovery and Its Importance

The reintroduction of the equant into the study of planetary motion was another step towards constructing a truly heliocentric astronomy, which would in turn lead Kepler to the discovery of the laws of planetary motion. To many of his contemporaries it looked like a step backward, but Kepler, convinced that it was a move in the right direction, was once again correct. One may say that reintroducing the equant was in no way original; already more than fourteen centuries before, Ptolemy had introduced this idea. Such a view betrays an extremely shallow perception of Kepler's move and a lack of knowledge of what exactly he did accomplish and how significant it was. The idea as such was not new. However, how he used and interpreted it rendered it original and productive.

Ptolemy had acknowledged the observed fact that the planets moved in their orbits with nonuniform speed. Wanting to be faithful to the ancient stipulation of uniform circular planetary motion, however, he too admitted that the planets moved in circles with uniform velocity, but with respect to an off-center point called the equant. For many astronomers this entailed a violation of the time-honored principle, because the uniform motion was no more around the center of the deferent of the planet. Copernicus, for instance, revolted against it, and to remedy this defect and restore uniform motion about the center of the deferent he introduced an epicyclet. Thus Copernicus had banished the equant from his astronomy and this he considered one of his great achievements. In clear opposition, Kepler's adoption of the equant is a remarkable testimony to his independent mind and uncanny scientific intuition. Despite all his great respect for Copernicus, Kepler never allowed himself to be imprisoned by his master's ideas.

In the MC Kepler gave a physical explanation for the equant.[28] He unhesitatingly admitted that others before him had introduced this idea. "However, the reason and the means are shown more clearly by these writings of our own. . . ."[29] He argued that the cause of variation of the planetary motions could be explained as a special case of the general theory proposed in chapter 20 of MC, where he had postulated the existence of a motive force in the sun, the strength of which varied inversely with distance from it. Whether a planet was accelerated or retarded depended, therefore, on its distance from the sun. "The path of the planet is eccentric, and it is slower when it is farther out, and swift when it is further in. For it was to explain this that Copernicus postulated epicycles, Ptolemy equants."[30] Kepler's equant circle was a consequence of this observation, because, according to it, the uniform motion of the planet could not be around the geometrical center of the circle since the sun on which the motion depended was at an off-center point. The center of uniform motion, therefore, must also be an off-center point, i.e., the equant. This meant that for Kepler the equant was not an artificial device, but a real consequence of a physical theory that endowed the sun with special characteristics. Ptolemy had accepted the equant without proof, or perhaps on faith, but Kepler found a reason for it. Hence he wrote: "Anyone who reads this passage on the equant will, I know, rejoice. For if the astronomers are surprised that Ptolemy assumed the same measure of the center of the equant without proof, some people will now be all the more surprised that there was an explanation for it, and Ptolemy did not suspect it, since he assumed the fact to be as it is, and as if by divine guidance arrived blindly at the proper destination."[31]

It now becomes clear that those who accused Kepler of retrogression in the case of the reintroduction of the equant failed to appreciate the originality and creativity involved in his move. Despite seeming similarities, his equant differed from Ptolemy's in radically significant ways. For Ptolemy this was an ad hoc artificial device to "save the phenomena" of observed nonuniform motion, whereas for Kepler the equant disclosed a real state of affairs: it revealed the fact that planets in reality executed nonuniform motion around the center of the orbit. Again, for the ancient astronomer, it was not a universal phenomenon, because the orbit of the earth (sun) did not have an equant, but for our astronomer no planet, not even the earth, was exempt from the rule of the equant. Furthermore, Ptolemy, being concerned only about "saving the phenomena," did not assign any real cause to the phenomenon. Already in the MC Kepler attempted to assign a real cause. Hence, at best, Ptolemy offered only a kinematics, whereas Kepler gave a dynamics. At best, Ptolemy only partially described what was observed, whereas Kepler both fully described and causally explained the phenomenon.

Kepler's original interpretation of the physical significance of the equant had a revolutionary consequence. It broke with the age-old tradition according to which uniform motion was considered the natural and only motion possible for heavenly bodies. Indeed, nonuniform motion was observed, but was relegated as something only apparent, since the only real motion had to be uniform. Kepler's interpretation exactly reversed this concept. According to him, nonuniform motion was the real motion, since it alone had a physical explanation, it alone had a physical cause. The uniform motion of planets, on the other hand, was only a pure theoretical concept since it had no basis in reality. His new interpretation of the equant had tremendous significance because it changed our concept of planets and planetary motion. Planets used to be defined as bodies executing uniform circular motion, and uniformity and circularity were two of the essential characteristics of planetary motion. Here Kepler dared to challenge the first idea. Perhaps this break with the age-old tradition made it less difficult for him to break with the second characteristic, circularity, later on.

Scientific or Empirical Ideas

In this "rediscovery" of the equant, observations and other empirical considerations played a role. For instance, Ptolemy introduced the device because he had observed that the motion of the planets was not uniform around the center of the orbit, but it could be made uniform only around an off-center equant point. Kepler had also noted that a planet's motion was faster when it moved closer to the sun and slower when away from it.

However, observations alone could not have led him to the point he actually reached, for at the time he reintroduced the equant, the observations available to him were not much better than what Ptolemy had. There is ample evidence that Kepler had made the "rediscovery" already in MC; for instance, he wrote to Herwart in 1600 that already in MC he had regretted that the earth alone did not have an equant.[32] Hence he made at least the initial "rediscovery" before he came to Tycho. Since the observations then available to Kepler were no better than those Ptolemy had had, there was no reason why he should have gone beyond the ancient astronomer, if the "rediscovery" was dependent on observations alone. But since it was not, far more than empirical considerations were involved. In fact, it seems that empirical considerations had only a rather minor significance here, because Kepler differed radically from Ptolemy and the difference could not be attributed to the availability of better observations. This is an instance where philosophical and theological ideas played extremely significant roles in helping Kepler make a crucially important move.

Philosophical Ideas

Generalization of the equant as a basic property of planetary orbit was one of Kepler's original insights. The equant became not just an ad hoc device for a few heavenly bodies, but a universal characteristic of all planetary orbits. Although later on he did give empirical proof, at the time of the "rediscovery" he did not furnish any such proof. Kepler could make such a generalization because he subscribed to the principle of generalizability, according to which whatever is true of one case must be true of other cases of the same kind.

The use of the principle of causality was also very conspicuous in reintroducing the equant. Kepler had noted a variation in the speed of the planets. The principle of causality argued that the variation should have a cause. He identified the cause with the *species* emanating from the sun. The equant, for Kepler, was the manifestation of this cause-effect relationship; the presence of this cause meant that the motion could be uniform not about the central point, but about some other point, viz., the center of the equant circle.

The principle of realism also operated in this context. Kepler believed that an astronomical theory should represent real motion. The uniform motion obtained by Copernicus's rejecting the equant was not real. And the Ptolemaic equant spoke of motion around an empty point. But Kepler asked: "How can any power residing in a non-body, flow out from that non-body into a planet?"[33] Through his original interpretation of the equant , he then linked the motion to a real body, the sun. There was another reason as well: Copernicus's rejection of the equant brought in new, unreal epicycles, and anything which had recourse to such unreal epicycles was revolting to Kepler's principle of realism. In this way the principle of realism opposed the Copernican idea and supported the Keplerian move.

Could these empirical and philosophical considerations alone have given Kepler sufficient motivation to make this "rediscovery"? We must remember that Kepler's move meant going against what his master Copernicus considered as one of his greatest achievements. It also meant sacrificing the symmetry that Copernicus claimed to have achieved and that Kepler himself valued so much.

The reintroduction of the equant also entailed staunch opposition from friends and foes alike: from friends because he opposed Copernicus, from foes because he seemed to be taking a step backward while he claimed to be marching ahead to build a new astronomy. For example, according to Tycho, the leading astronomer of the day, the idea of the equant was far more absurd and abhorrent than that of a moving earth:

And his [Copernicus's] apparently absurd opinion that the earth revolves does not obstruct this estimate, because a circular motion designed to go on uniformly about another point than the very center of the circle, as actually found in the Ptolemaic hypothesis of all the planets except that of the sun, offends against the very basic principles of our discipline in a far more absurd and intolerant way than does the attributing to the earth one motion or another which, being a natural motion, turns out to be imperceptible.[34]

According to the Dane, the circular motion should be uniform around the center of the circle, not around an off-center point, as the equant theory claimed. These objections could not have been taken lightly, especially when Kepler's empirical reasons were very insufficient. I think that a powerful additional reason that impelled him to this idea, despite the objections, was that the move helped him build a truly heliocentric system.

Religious Ideas

As we shall discuss in chapter 8, reintroducing the equant helped Kepler make the sun not just the geometrical center but also the true dynamic center of the universe. In fact, the "rediscovery" was the first step towards that change of view about the sun. Thus one of the main reasons for the "rediscovery" was that it ensured a preeminent position and role for the sun. We have already seen the religious roots of this belief: Kepler believed that this innovation would make it possible to construct a God-centered astronomy. Since his life-long ambition was to build up a God-centered view of the world, he had a powerful fascination for any step that contributed towards that goal.

Another way that religion played a role in this "rediscovery" was in Kepler's having apparently been convinced that the presence of the equant was, as it were, divinely certified. In speaking of Ptolemy's discovery, Kepler remarked that the ancient astronomer "by divine guidance arrived" at this idea.[35] Hence he believed that his own decision to reintroduce the equant was guided by God, a consideration with a powerful grip.

Kepler's attempt to create a truly heliocentric view of the world can be summarized in three short propositions: (1) There is one and only one point to which all aspects of planetary motion (distance, tilt or inclination of the orbits, point of intersection of orbital planes, center of the orbit, cause of planetary motion and its variation) can be referred. (2) That point cannot be empty, it can only be a certain physical body. (3) The only body suitable to occupy this point is the sun. Establishing a heliocentric system meant establishing all three propositions. It is quite clear that in the steps discussed in this chapter these propositions are involved. They bring out most conspicuously the fact that the creation of the truly heliocentric system required not only empirical considerations but also philosophical and religious ideas. For it is quite clear that for establishing the first proposition, empirical reasons were crucial. For establishing the second, Kepler needed to consider especially several of his philosophical principles. Furthermore, to establish the third, he had to have recourse to his religious beliefs. It follows, therefore, that his creation of the truly heliocentric system was a striking example of the threefold interaction of empirical, philosophical, and religious principles.

6

THE VICARIOUS HYPOTHESIS
AND ITS FAILURE

THE DISCOVERY

Having finished the necessary preliminaries, i.e., having set the correct point of reference, determined the nature and value of the inclination of the orbits to the ecliptic, and established the need for the equant for the orbit, Kepler was ready to develop his theory of Mars. Accordingly, he selected twelve observations at oppositions—ten from Tycho's and two from his own. Kepler also verified some of these observations by comparing them with data obtained by Fabricius in East Friesland.

Kepler wanted to determine the orbit of Mars. These elements—the position of the line of apsides (longitude at aphelion), the eccentricity, the period, the epoch, and the size of the orbit—would determine the orbit, since at this stage, like all other astronomers of the day, he believed in the inviolability of the principle of circularity. However, unlike most of his colleagues, he refused to subscribe to Ptolemy's bisection of the eccentricity according to which the center of the orbit of Mars exactly bisected the distance between the centers of the equant and the sun (earth). Kepler considered such a bisection totally unwarranted and a source of errors and imperfections in the

162

traditional astronomy. Of course, later investigation would show that the bisection was correct, but for the time being he refused to assume it. He wanted to investigate afresh on the basis of observational data the ratio in which the distance between the centers of the sun and the equant was to be divided.

Having challenged Ptolemy's bisection of the eccentricity even before coming to Tycho in 1600, Kepler was delighted to note that the Dane had divided it in a different ratio. Although Copernicus had great respect for Ptolemy, he, too, doubted the bisection;[1] besides, since the Canon claimed to have abolished the equant, there was no Ptolemaic kind of bisection of eccentricity in his system.

In figure 5, GFQDHP is the circular orbit of Mars with center at O (see figure 5). PQ is the diameter representing the line of apsides. Assume that the observations are made from the point S. E is the center of the equant. Let D, H, G, and F be the positions corresponding to the four oppositions.[2] From observations we can calculate the angles within the lines SD, SF, SG, and SH. The problem now is to determine angles PEG (a) and PSG (b), i.e., the mean and true anomalies, in such a way that the points D, F, G, and H, may be found in the circumference of the circle whose center O will lie on SE, and to find the distance of O from S and E.

Kepler used the method of false position to carry out this program. He had to assume PEG and PSG and then see what followed from this assumption. If the assumed values led to contradictory or inconsistent consequences, then the assumptions were wrong and new ones would have to be made.

Observations can give SD, SF, SG, and SH, the directions of the heliocentric longitudes. Also the angles GSF, FSD, DSH, and HSG can be found from observations. Again, the angles GEF, FED, DEH, and HEG can be obtained from calculation, since the period of revolution and hence mean motion of Mars are known.

In triangles ESG, ESF, ESD, and ESH, ES, i.e., the distance between the equant and the sun, is the common base.

Now angle FES = 180 − (a + angle GEF) = g
Angle FSE = GSF + b = d
Angle EFS = 180 − (g + d) = t
In triangle ESF all the three angles are known.
Therefore ES/Sin t = EF/Sin d = SF/Sin g.
Therefore SF = ES Sin g /Sin t.

FIGURE 5

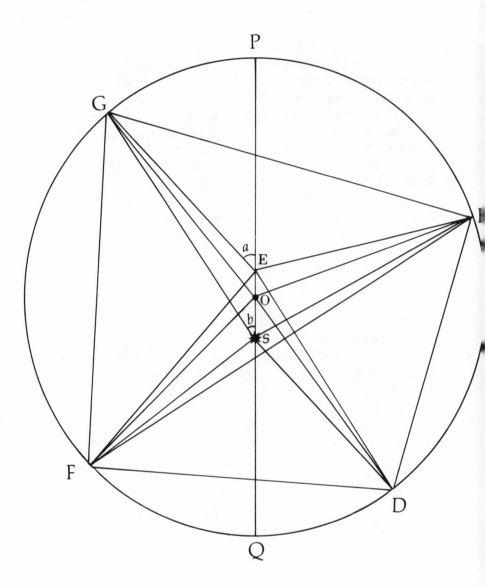

Similarly SD, SH, SG can be obtained in terms of ES. Hence the respective distances of Mars from the sun can be determined in terms of ES.

Since we know SF, SG, and angle GSF,
SinGSF/FG = SinGFS/SG = SinFGS/SF
Therefore SinFGS = SF/FG SinGSF

FG can be found.
$FG^2 = SF^2 + SG^2 - 2SF \times SG$ CosGSF.
Therefore angle FGS = Sin^{-1} (SF/FG SinGSF).

Similarly angle SGH can be found. The angle FGH can be obtained. Similarly angle FDH also can be found. If angle FGH + angle FDH = angle DHG + angle DFG = 180, we have a circle. If this is true (or at least nearly true), then we have a circular orbit for the planet Mars. If not, we need to readjust the original assumption and carry out the procedure again.

The next step is to locate O and to determine the distances OE and OS. We have the lengths of FG, FD, DH and HG in terms of ES. Consider the isosceles triangle FOG.

We know FG.
Angle FOG = 2 angle FHG
= 2 (angle SHG + angle SHF).
These angles can be known.
But angle OGF = angle OFG = 1/2 (180 − angle FOG).
In triangle FOG, OG/SinOFG = OF/SinOGF = FG/SinFOG.
Therefore radii OG and OF can be known in terms of FG
which in turn can be known in terms of ES.

Consider triangle OGS.
Angle OGS = angle OGF − angle FGS.
Therefore OS can be found.
Also we can find angle OSG.

If O is on ES, then OS will fall on ES, in which case angle OSG = angle PSG. If this turns out to be the case, the work is done. If not, the assumed values of *a* and *b* will have to be altered, and the whole process repeated.

Kepler had to do this laborious operation seventy times over. The results obtained seemed to have confirmed his intuition that the bisection of the eccentricity was unfounded. Calculations showed that if the radius of the orbit was taken as 100,000, then the eccentricity would be divided in the ratio of 11332 to 7232. Thus he seemed to have accomplished the task set before him of improving the old astronomy. This hypothesis of the circular orbit of Mars he called the "vicarious hypothesis" (*hypothesis vicaria*) and he used it until he discovered the correct orbit.

He now wanted to verify this theory by comparing the positions it gave with the positions actually observed. He was greatly heartened to note that the longitude determination verified his theory with unprecedented accuracy. The error never exceeded 2 minutes and 12 seconds; traditionally, an error of 10 minutes had been perfectly acceptable,[3] so this accuracy was very encouraging and he rightly exulted in his success.

However, he was soon to learn that there was no easy victory over Mars. The determination of latitude at opposition turned out to be disastrous, showing considerable discrepancy between theory and observation. But he noticed that if he accepted the bisection of the eccentricity, he could get a good result for the latitude that would differ only by 310 parts in terms of the semidiameter of the planet's orbit, being taken as 100,000. This led him to argue that Ptolemy's bisection was not a mere simplifying assumption but must have been a conclusion arrived at after repeated trials on the basis of observational data.

Not even the acceptance of bisection, however, could ensure Kepler's final victory. He could get satisfactory agreement between observed and calculated positions as long as he considered the apsides and positions 90 degrees away from the apsides. But when intermediate positions of 45 and 135 degrees were considered, the discrepancy mounted to the order of 8 minutes, a broader margin of error than Kepler was willing to accept (despite the fact that for Ptolemy such a discrepancy was not significant). Kepler's refusal was the turning point in his career as an astronomer.

Perhaps in no other instance is his emphasis on observation or empirical inquiry more conspicuous than in the discovery of the vicarious hypothesis. According to Koyré, this reveals most strikingly Kepler's "Tychonism" or "empiricism." The derivation of the vicarious hypothesis showed that Kepler really meant what he had said: "Without proper experiment I conclude nothing." Again and again, we find him making calculations and checking them with observations. Neither hard work nor loss of time could stanch his determination to derive a theory on the basis of accurate empirical data.

Kepler's emphasis on observations to arrive at the correct astronomical theory was even more striking in his refusal to accept the time-honored bisection of the eccentricity. Several considerations should have encouraged him to follow tradition: the authority of Ptolemy, the practice of other astronomers, and the simplification of procedure and consequent gain in time and effort. A more significant consideration could have been the symmetry involved in such a bisection, for the harmony which revealed itself in symmetry was very appealing to Kepler. What would have been more harmonious and symmetrical than having the eccentricity equally divided by the center of the orbit? But he refused to be influenced by any such considerations and wanted observational data to be the principal judge on the matter. This change in principle and method attested to a Kepler far from the MC, where he had boasted of having derived astronomical truths a priori.

The discovery of the vicarious hypothesis seemed to have brought Kepler the empirical scientist to the forefront. But did this mean that philosophical and religious ideas had shifted to the background, or even to their burial ground? Some historians, like Dreyer, probably believed that this was the case. According to them, as Kepler's thoughts and works matured, he distanced himself more and more from his philosophical and religious ideas and commitments. They see a Comtean transition or "development" taking place in Kepler: As his thoughts developed, he shed his religious views to enter the metaphysical stage. Then he discarded this second phase to move into the mature stage of empiricism. However, instead of having been a paradigm of the Comtean scheme of "development," Kepler was really a persuasive counterexample. His religious and philosophical views continued to influence him to the end, even during the heyday of AN. This point is clear in his reaction to the failure of the vicarious hypothesis.

THE TURNING POINT IN
KEPLER'S WORK IN ASTRONOMY

With the failure of the vicarious hypothesis Kepler's work reached a critical stage, a moment of decision—a decision which would affect his whole future career. To his great dismay and disappointment he found an error of 8 minutes. The choice before him was whether to take the same path that his predecessors and contemporaries had taken[4] or to reject the tradition-honored path and plunge into the unknown, uncharted path. It was an excruciatingly difficult, but crucial, choice. Indeed, the whole future of astronomy depended on it. As he himself said, the decision concerning these 8 minutes "provided the foundation, and pointed the way to the reformation of the whole of astronomy."[5] A margin of error higher than he was willing to accept forced him to reconsider the principles to which he had hitherto subscribed.

REASONS FOR KEPLER'S DECISION
TO START ANEW

Scientific or Empirical Ideas

We know that Kepler made the right choice, which won for him the rightful honor of the title, founder of modern astronomy. Now what induced him to make the right choice, despite its extreme difficulty and crucial importance? Observations and accurate agreement between theory and observations he considered absolutely critical. He looked for a theory which would agree with observation in all cases and under all circumstances. He was fully aware of the labor involved in the study. Thus he wrote to Longomontanus in the beginning of 1605: "You warn me that by seeking physical causes, I might upset the heavens again. By my Christ, if I had wished to doubt the 8 minutes, I would have been able to omit more than three times as much effort as I put into the year 1604. Thus you know how diligently I labored, that from the observations I might be meticulously precise."[6]

Kepler's refusal to neglect the error of 8 minutes left him with two options: either the orbit was not circular, or, if it was, it had within it no fixed point around which the planet could perform uniform motion—if there was such a point at all, it had to be librating upwards and downwards along the line of apsides.[7] Kepler rejected both these options and decided to start afresh. As he himself related, this was the beginning of his new astronomy.

Several writers like Dreyer and Small seem to believe that empirical reasons alone were sufficient to induce Kepler to refuse to accept the error of 8 minutes. There is no doubt that he insisted on accuracy. And he wanted the same theory to account for both longitude and latitude, although previous astronomers would have been quite content with a different theory for each of these two parameters. The critical questions are these: Why did Kepler insist so much on these points? Why did he consider 8 minutes too far off the mark? After all, other astronomers had also emphasized accuracy, but an error of 10 minutes was not too much for most of them. Why did Kepler harp on so high an accuracy, although he knew that his insistence meant rejecting something for which he had worked so assiduously and tirelessly for four long years? Why did he insist that the theory be true of all possible positions of the planet and for all parameters?

Philosophical Ideas

An adequate answer to questions about Kepler's decision to start anew should take us to his philosophical and religious beliefs. A number of Keplerian philosophical principles were actively involved: realism, geometrizability, precision, generalizability, simplicity, and economy.

Unlike Plato, Kepler believed that our universe is real, through and through. Any theory explaining natural phenomena must be amenable to observational and other empirical tests. The influence of the principle of realism was evident as well in his insistence that the same theory should account for both longitude and latitude. Previous astronomers had been interested in "saving the phenomena" only, and so they did not hesitate to employ two different models. But for Kepler a scientific theory should be a true representation of reality and hence should explain different aspects of that reality.

The principle of geometrizability was another belief Kepler adhered to faithfully. For him, nature is mathematical in the sense that mathematical laws were used in the creation of the universe. This relationship between the material universe and geometry was well expressed in his statement "Ubi materia, ibi geometria," "Where there is matter, there is geometry." Wherever matter is involved, there geometry also is applied. Obviously, theories of nature must be amenable to geometrical investigation. Hence we see him applying generously geometrical theorems and arguments in the development of the vicarious theory.

This geometrical nature of matter further manifested itself through his search for precision and accuracy. He demanded precise agreement between theory and observational data. Although other astronomers were willing to strive for great accuracy, it was not clear to everyone that the universe itself was precise. For instance, Simplicius in the *Dialogues Concerning Two World Systems* argues that nature does not obey the laws precisely. Salviati seems to have agreed with that view, to which even Galileo appears to have subscribed. Kepler, however, refused to go along with it. Had he remained a strict orthodox Platonist, he would not have been so insistent on accurate agreement with observational data. Thus here the principle of precision, which we have seen as a special case of the principle of geometrizability, came into play. He wanted a better accuracy than 8 minutes. Of course he was aware that instrumental and human errors would never allow 100 percent accuracy; he was not pursuing such an impossible goal. He admitted that there was a limit to the knowledge that could be accessible and attainable to human beings. As he told Herwart, he had no difficulty in accepting such a limitation. In this case, however, he knew that Tycho's instruments and techniques could yield an accuracy up to 2 minutes. He wanted that kind of accuracy.

Generalizability also was operative in the discovery and rejection of the vicarious hypothesis. The principle demanded that theory agree with observation at all positions of the planetary orbit, not at a few privileged positions of the aphelion and perihelion. The discovery that this principle could not be satisfied by the vicarious hypothesis was certainly one of the reasons for its rejection.

The principles of simplicity and of economy also played a remarkable role in his decision to start afresh, which in turn layed the foundation for the new astronomy. As we have already seen, his refusal to ignore the error of 8 minutes left him in a dilemma: to opt either for a noncircular orbit or for a librating point inside the orbit around which the planet moved uniformly. Considerations of simplicity ruled out a noncircular orbit, since a circular orbit is the paradigm of simple natural motion. Referring to the librating equant point, he remarked: "I do not see how that can be reconciled with natural reason."[8] It could not be consonant with natural reason because it violated the principles of simplicity and economy. A system where the planets had to execute uniform motion around a central point which continually moved up and down could hardly be considered a simple arrangement. Such a system could not be in agreement with the principle of economy either, since to ascertain exactly how far to librate so as to sustain uniform motion of the planet, much calculation would have to be made.

Religious Ideas

Our discussion makes clear that at this crucial turning point of Kepler's work several of his philosophical principles were actively involved. In part 1, we have seen that for Kepler these principles were deeply rooted in his religious beliefs, were the direct outcome of his religious beliefs, especially of his view of God.[9] How his religious beliefs influenced him in rejecting the vicarious hypothesis can be illustrated by his belief in God the geometer. Kepler believed that the geometer God created the universe according to the laws of geometry and made humans with the ability to understand mathematics. Given these ideas, accepting an error of 8 minutes would lead to one of two conclusions: either God did not create the universe with geometrical accuracy, or else humans were not capable of attaining a better accuracy and so had to be content with inaccurate results. Both conclusions were unacceptable. The first must be rejected because it would be an insult to God. It would violate another characteristic of God, viz., God's being the best and wisest creator. God would be reduced to a poor artisan who could not make a perfect universe. Kepler's idea of God was not Newton's. Newton could tolerate the idea of a God who made an imperfect world that needed God's periodic intervention. No, the God of Kepler was no "repairman God," but, rather, the Leibnizian God (or, more accurately, the Leibnizian God was the Keplerian God) who created the most perfect world possible and who operated in accordance with the principle of sufficient reason. The second alternative, too—that humans could not attain accuracy—was unacceptable, because Tycho had shown that an accuracy up to 2 minutes could be achieved in this astronomy. Far better results than an 8 or 10 degree margin of error were within reach of human inquiry. Therefore the error in the vicarious hypothesis could not be accepted.

Another religious consideration also played a significant role in helping Kepler make the right decision. He believed that Tycho's observational data were a special gift from God and so had to be taken with utmost reverence and seriousness; they had to be used in the best way possible. In his own words: "Seeing that Divine Goodness has given us in Tycho Brahe a most diligent observer whose observations have revealed the error of 8 minutes in Ptolemy's calculation, it is only right that we should gratefully accept this gift from God, and put it to the best use. By means of this gift let us strive to discover ultimately the true nature of celestial motions."[10] According-ing to Kepler, disregarding the error meant failing to appreciate the full merit and worth of God's gift, because these very observations had revealed the error. Failing to give due credit to God's gift was tantamount to an insult. Kepler had a more positive consideration as well: he took this occasion as God's beckoning humans to explore at length the true nature of celestial motions.

Kepler's acceptance of the full weight of the failure of the vicarious hypothesis and his consequent decision to start anew were not the results of empirical considerations alone. Philosophical and religious ideas also had significant roles to play. All three factors contributed to bringing into range the all-important turning point in his astro-nomical work.

7

THE FINAL BREAK WITH GEOCENTRISM

Not only did Copernicus fail to give a genuinely heliocentric astronomy, but he also could not extricate himself fully and definitively
from geocentrism. His system still retained its vestiges. True, according to his theory the earth was not the center of the universe around
which all the heavenly bodies revolved. Yet the earth continued to
enjoy a privileged position. Despite his revolutionary thinking, the
spirit of Aristotelian geocentrism persistently haunted Copernicus to
the end. For instance, he had taken the center of the orbit of the earth
as the point of reference, which obviously gave the earth a special
status among the other planets. Moreover, in his system the earth's
sphere, unlike the spheres of the other planets, had no thickness, since
its path was everywhere equidistant from the mean sun. Further, in
the Ptolemaic system, the planes of the planetary system had been
so constructed that they intersected at the center of the earth, and
Copernicus preserved this privilege location of the earth, though in
a new form (the orbital planes intersected at the center of the earth's
orbit). Finally, in the consideration of eccentricity, he treated the
earth in a different way from all the other planets. In the DR the
eccentricities of all the planets except that of the earth were measured
from the center of the earth's orbit, whereas that of the earth was
measured from the sun. Clearly, just as Copernicus's acceptance of

heliocentrism was incomplete, his rejection of geocentrism also was far from total.

Just as it was Kepler who seated the sun really at the center of the universe, it was Kepler who really unseated the earth from the center of the astronomical world. This unseating of the earth was yet another proof of the unique and divine-like status of the sun in his system. This chapter briefly discusses Kepler's study of the earth, which led to the establishment of the fact that the earth, astronomically speaking, was just one of the planets.

Kepler had another reason to take up the study of the motion of the earth. The failure of the vicarious hypothesis forced him to subject his methods and ideas to strict reexamination. In that light he decided to alter the traditional strategy of studying the first inequality first. Up to now, like all his predecessors, he had been concentrating on the study of the first inequality. Now he decided to focus his attention on the second inequality. Explanation as to how this inequality came about varied from one astronomical system to another. According to Ptolemy, the inequality was caused by the motion of a planet on its epicycle. Tycho Brahe explained it in terms of the sun's motion around the earth. Copernicus, on the other hand, argued that it arose from the motion of the earth, because of the projection of the earth's orbital motion onto the other planets. Naturally, Kepler subscribed to the Copernican view. He considered the study of this inequality most fundamental because we make the observations from the earth. Since the distance of the earth from the sun was extremely important in astronomical investigations, necessarily the orbit and eccentricity of the earth's motion had to be determined accurately.

THE IMPORTANCE OF
THE TOTAL REJECTION OF GEOCENTRISM

Study of the orbit of the terrestrial motion, especially the establishment of the fact that the earth is just another planet characterized by an eccentric orbit and equant, Kepler considered the key to his astronomy. As late as 1621 in a note to the second edition of MC he wrote:

But in my *Commentaries on Mars* I have made this [that the earth needed an equant] one of the chief features of the book, and have laid it like a cornerstone at the foundation. Indeed I deservedly called the key to astronomy the fact, which I have demonstrated clearly from the actual motions of Mars, that the annual motion either of the sun or of the earth is controlled by a different center from the equant, and that the eccentricity of its orbit is only half the eccentricity believed by the authorities.[1]

Hence for him the earth was very much one among the planets.

It may be noted that Kepler, an ardent Copernican, could not shake himself fully free of the spirit of his master. In some ways he also accorded a certain special importance to the earth. Thus he ascribed to the earth a certain nobility; our earth was "the pinnacle and pattern of the whole universe, and therefore the most important of the moving stars [planets]."[2] He considered it most appropriate that in his five solids theory of the solar system the earth should occupy the place between the two kinds of solids, i.e., between the primaries (cube, tetrahedron, and dodecahedron) and the two secondaries (the octahedron and the icosahedron).[3] Aiton points out that this "geocentrism of importance" postulated by Kepler was similar to the "heliocentrism of importance" that Renaissance Platonism attributed to the median position of the sun between the earth and the fixed stars.[4] In his letter to Herwart in 1599 Kepler wrote: "God has called it [the earth] an exception adorning it alone with the orb of the moon."[5] But he attached only a positional significance to the earth. He denied it any special astronomical significance, in the sense that he considered it simply one of the planets in every way. He was the first to treat the earth fully as a planet—a noble and extremely important one, yet a planet.

That the earth should be considered a planet in every way came to him quite early in his career. In fact, he stated that already in MC he had suspected it. We read in AN[6] that when he was discussing chapter 22 in MC he had conceived the suspicion that the earth should have an equant. In his letter to Herwart in 1600 he wrote: "Moreover, at the end of my MC I have spoken of the equant of the sun, expressing my regret that it was denied to the sun (or earth) alone. . . ."[7] The same idea he mentioned to Magini in Bologna a year later: "In another chapter in this book [MC] I warned about the theory of the sun (earth) that it is not endowed with an equant by the practitioners [artificibus], but remains with a single eccentricity. This I suspected of falsehood."[8]

It follows that Kepler believed all along that the earth had an equant like all the other planets and that the belief of his predecessors and contemporaries that the earth's equant point and the center of its orbit were coincident was unfounded and incorrect. The failure of the vicarious hypothesis and his subsequent determination to start afresh and reexamine the traditional principles gave him an apt occasion to submit his suspicions to critical study.

In figure 6, QCRD is the earth's orbit. DC is the line of earth's apsides. S is the sun. O is the center of the orbit. E is the center of uniform motion.[9] When the planet is in T, and when the observations are made from the earth in the opposite quadratures, Q and R, the angle of parallax, QTE and RTE will be the same whether E and O are coincident or not. But if the planet were at F, i.e., on a line perpendicular to CD, and if it were observed from D and C, the angle of parallax DFE and CFE would be equal only if E and O coincided. Since the vicarious hypothesis can yield the longitude of the position of Mars, with the help of this hypothesis the angles DFE and CFE can be determined.[10] On the basis of this theory Kepler examined Tycho's observations and found just what he had anticipated, thereby proving that the equant and the center of the orbit did not coincide in the case of the earth.

Kepler conducted several different studies using different techniques. All confirmed his suspicion. Koyré describes one of the studies which was perhaps the most striking. Here Kepler decided to transport the observer to Mars and make all the observations from this planet. The method his genius devised to achieve this feat was to use only those observations taken when Mars was in the same place on its orbit. Thus if the individual observations were separated by an interval equal to the period of revolution of Mars (687 days), he could get the effect desired. The investigation using this ingenious method showed conclusively that the earth's eccentricity was nearly bisected. He found that the equant and the sun were approximately equidistant from the center of the orbit. The value of the distance turned out to be approximately half of the eccentricity ascribed to it by Tycho.

The discovery that the earth, like the other planets, had an equant was an essential part of Kepler's original work in astronomy. In

FIGURE 6

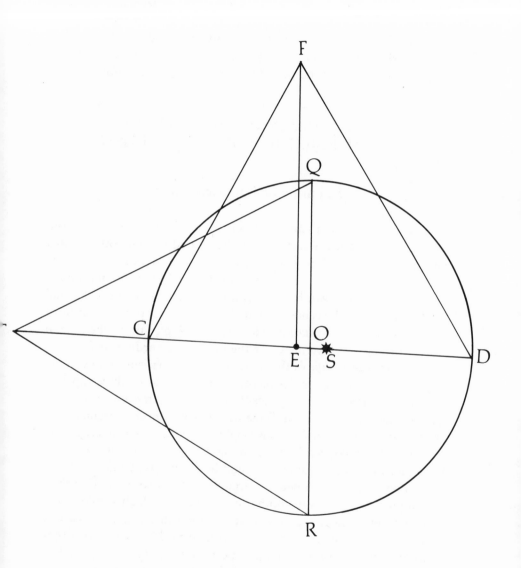

fact, it played a crucial role in his eventual discovery of the laws of astronomy. As he himself put it, it was the "cornerstone at the foundation" of his astronomy. This move meant depriving the earth of the last vestiges of special importance, from an astronomical point of view. Copernicus and his other followers would, no doubt, have found it shocking to rid the earth of special significance and hence did not even consider the possibility of making such a move.

FACTORS RESPONSIBLE FOR THE
TOTAL REJECTION OF GEOCENTRISM

Scientific or Empirical Ideas

As I have already discussed, observations were important in convincing Kepler that the earth should be treated like any other planet and that it, too, had an equant. But empirical investigations alone could not have enabled him to reach this conclusion. In fact, observations seemed to have had the opposite effect on Tycho, who had made several which seemed to indicate that the Copernican orbit of the earth did show variation. He wrote to Kepler that "the annual orb according to Copernicus, or the epicycle according to Ptolemy, does not appear always of the same size, in comparisons made to the eccentric, but introduces a perceptible alteration in all three superior planets, so much so that for Mars the angle of difference reaches 1 degree 45 minutes."[11] For the Dane this was yet another argument to show that Copernicanism was wrongheaded and hence unacceptable. On the other hand, Kepler used the same observation, not to refute Copernicanism, but to improve it and lead astronomy ahead. But he could not have done so without arguments from sources other than the empirical. There was another consideration which supports the view that much more than observations and other empirical reasons was involved in this case. It is quite clear from his own writing that many of the observational data, especially those of Tycho, reached Kepler only after he had suspected the need for an equant for the earth. For example, as I have pointed out earlier, already in the MC he had expressed this view. Hence it is clear that the origin of this idea cannot be attributed to empirical factors alone.

Philosophical Ideas

Kepler's philosophical principles of harmony, generalizability or universalizability, and simplicity contributed in significant ways to his rejection of geocentrism. Leaving the earth as an exception would have undermined the harmonious arrangement of the planetary system. All the other planets had an equant, while one planet alone did not—and for no apparent reason. Furthermore, he believed in the principle of universalizability, which meant that all planets, as planets, behaved in the same way and had similar properties. According to the old view, what was applicable to all the other planets was not so in the case of the earth. This obviously challenged the principle of universalizability. The old Copernican view would go against the principle of simplicity as well. One of the main characteristics of systematic simplicity is the ability to explain a maximum number of phenomena with a minimum number of explanatory factors. In the old model, since the earth was an exception, the general law could not be used in its case, which meant that astronomers needed to bring in an extra law or explanatory factor. On the other hand, in Kepler's model, one and the same general law was applicable to all the planets and even to the moon.

Religious Ideas

For Kepler the equant was not merely a mathematical device, introduced to simplify calculations or to "save the phenomena." The equant had a physical significance: it argued for the presence of an impelling force. For if the equant was the point around which the planets moved with uniform velocity, and this point and the sun were placed symmetrically on the line of apsides with respect to the center of the deferent, and if the maximum and minimum speeds were obtained at perihelion and aphelion respectively, then clearly the motion of the planet, at least at the apsides, was directly related to the position of the sun. If the earth did not have an equant, then direct relationship between the sun and its motion would be in doubt. Since this relationship implied a dependence of the planet on the sun, or at least on its position, the absence of the equant for the earth could imply the absence of such a dependence. On the other hand, if it could be shown that the earth also had an equant, then the earth also could be considered to be under the influence of the sun and thus the supremacy and unique importance of the sun would remain intact. The unique importance Kepler accorded the sun also had a role to play in this discovery.

Why did Kepler insist on preserving the unique position of the sun from the least bit of doubt? Why did he emphasize the principle of universalizability and simplicity? It may be noted that Copernicus had also stressed the importance of the sun, yet he never made the kind of moves Kepler did. Kepler differed from Copernicus because for him the sun was far more than just an important heavenly body. The sun was the visible image of the invisible God the Father, the creator and master of all things. Just as God had supreme dominion over everything, the sun should have dominion over all planetary bodies. Nothing could be exempted from its influence. If the earth had an equant, it meant that its speed of motion was dependent on the position of the sun, and so the earth, too, was dependent on the sun. God's laws are universal, applicable to all beings. Similarly, laws involving the position and operation of the sun must also be applicable to all planetary bodies. Thus Kepler's religious views helped him reach the conclusion that the earth had an equant.

8

THE DISCOVERY OF THE SECOND LAW

The circumstances surrounding the discovery of original laws of science are often very strange and unpredictable. This seems to have been especially true of Kepler's process of arriving at the second law, which says that the velocity of a planet varies in such a way that its radius vector sweeps equal areas in equal times. First of all, although called the second law, it was in actuality discovered first. Kepler proposed it as a postulate in the process of his discovery of the first law; after he had established the first law, he gave a formal proof for the area law.

The discovery of the second law was intimately related to Kepler's attempt to make the sun the dynamic center of the universe and to build up a dynamics of the heavens. Kepler's great ambition was to create a true heavenly physics, as opposed to a heavenly metaphysics or even a heavenly mathematics. To be sure, metaphysics and mathematics had a place in his study of astronomy, but he wanted more: he wanted a system that would give a physical explanation of heavenly phenomena. He wanted to know why the planets moved the way they did and what the laws governing that motion were.

Although the issue of fashioning a real celestial physics would seem completely natural for any astronomer today, it was not a guiding concern in Kepler's time. The mathematical astronomy of Ptolemy had no interest in the investigation of the causes of planetary motion since the circles, epicycles, and spheres used by the system were pure computational devices bereft of any real physical significance. As Koyré puts it, "celestial motion belongs to the realm of pure kinematics: astronomy is one thing, physics is something quite different."[1] Nor did the advent of Copernicanism, despite all the revolutionary characteristics that have been attributed to it, change astronomy in any significant way. Being still imprisoned in the celestial spheres of Aristotle, Copernicus could hardly see any dynamic problems with regard to planetary motion; he believed that planets were embedded in the crystalline spheres which revolved by their very nature, by the fact that they had a spherical shape. Tycho, after repeated and careful investigation of the nova of 1572 and the comet of 1577, claimed to have demolished the long-standing belief about the reality of the solid spheres,[2] because these two bodies were found to be superlunar. Their sudden appearance showed that the superlunar space was penetrable and alterable. In addition to the comet and the nova, he had observed at least six more comets before the end of the year 1590, e.g., the comets of 1580, 1582, 1585, and 1590. The destruction of the solid spheres should have haunted Tycho's mind with serious questions about how the planets moved. In place of the solid theory, he seemed to have subscribed to a fluid theory of the heavens, and so some kind of vortex mechanism of planetary motion was to be expected. However, according to Koyré, "Tycho wavered between belief in planetary spirits and a purely computational attitude."[3]

On the other hand, the destruction of the solid spheres mushroomed a whole series of questions in Kepler's mind. How are the planets supported in the heavens? What accounted for their regularity? And the all important question: "A quo moventur planetae" ("What is it that makes the planets move")?

Another, related problem was the study of the actual path of a planet around a central body, a sort of study that was also something unheard of in Kepler's time. To quote Koyré once more, "a study of the orbit, and a determination of the distances of a planet from the sun were radical innovations."[4] Pre-Copernican astronomy was interested primarily in the angular distances. In fact, Koyré goes on to say that the orbit of a planet did not have any real existence as far as this astronomy was concerned, in the sense that it interested no one.[5]

It was in the process of this radically new project of building up a dynamics of the heavens, or a heavenly physics, that Kepler came to discover the second law of planetary motion. The discovery of this law can be divided into four parts: (A) the establishment of the distance law; (B) the establishment of the cause of the planetary motion, or his attempt to give a physical explanation for the distance law; (C) studying the actual path of the planet with the help of parts (A) and (B); (D) discovery of the area law as a method of simplification for part (C).

ESTABLISHING THE DISTANCE LAW

The Origin and Development of the Distance Law

Long ago, Ptolemy had noticed that the motion of the planet on the deferent varied according to the distance of the moving point, i.e., the center of the epicycle on which the planet moved, from the *punctum aequans*. The further the point, the faster it moved. Later on, Copernicus and Tycho also acknowledged such a variation of the angular velocity of planets with distance when they ascribed epicycles to the superior planets and hypocycles to the inferior planets. However, neither Ptolemy nor Copernicus presented any definite law governing the dependence of the speed of a planet on its distance.[6] Kepler, on the other hand, saw a dynamic physical law operative here. Expressed in modern terms, the law's geometric or kinematic aspect ran as follows: "The velocity of a planet in its orbit is inversely proportional to its distance from the body about which it revolves."[7] The dynamic aspect Kepler enunciated as follows: the influence (force) of the central body (the sun) on the planet varies inversely as the distance between them.

Through a geometrical method and using certain approximations, Kepler proved this law.[8] He showed that the law was applicable, not only to arcs at the apsides, but also to arcs at other positions of the orbit, although the proof for the second part was less geometrical. Since the farther a planet was, the slower its velocity, it followed that the time (*mora*) during which a planet remained on a given arc of its eccentric was directly proportional to its distance from the sun.

At first it may appear that the Ptolemaic idea and the Keplerian formulation were the same. In fact, in the beginning Kepler himself thought that the Ptolemaic idea was the counterpart of his own, because he mistakenly thought that the orbital velocity of the planet was inversely proportional to the distance. Later, in the *Epitome* he gave the correct law that the velocity perpendicular to the radius vector is inversely proportional to the distance from the sun.[9] At the apsides, since the orbital velocity and the velocity perpendicular to the radius vector coincide, in practice it did not make any difference which velocity we took, but that was not the case with other points of the orbit. So actually the Ptolemaic and the Keplerian ideas were not equivalent. Here a mistaken perception seemed to have put Kepler on the right track. Other considerations as well showed that the Ptolemaic and Keplerian ideas were not the same. The former, being applicable only to the arcs at the apsides, was particularistic. Kepler considered his law general, applicable to all the arcs of the orbit. Moreover, the old view expressed a relationship that involved two empty points: the center of the epicycle and the center of the equant. No physical or causal relationship could exist between two such points.[10] On the other hand, for Kepler, the relationship involved the central body and the body of the planet; one could talk of a causal relationship, for real bodies were involved and the law had a physical significance.

Factors Responsible for the Discovery of the Distance Law

SCIENTIFIC OR EMPIRICAL IDEAS

Observations played a crucial role in the discovery[11] of the distance law. The ideas of both Ptolemy and Copernicus seemed to have been based on observations. Since Kepler, in presenting his version of the distance law, did borrow from their ideas, we can confidently say that observations were involved in this discovery. Kepler's own statements also support the view that observational data and his reflection on them were crucial. For instance, in MC he openly referred to this idea immediately after he had studied the variation of velocity of the planets. He noticed that the more distant the planets were, the slower they moved.[12]

PHILOSOPHICAL IDEAS

Observations alone were not adequate to establish this law, as is quite clear from the fact that Kepler did not attempt to establish it by presenting empirical data alone, rather he gave a long geometrical proof. Here the principle of geometrizability played a pivotal role. The use of geometry was not based on mere pragmatic considerations. For him geometry was far more than an effective tool for computational purposes. The geometrical principle was deeply rooted in him and was one of the principal factors forming his worldview.

I have mentioned that for Ptolemy the dynamic relationship was applicable only for the apsides. Kepler generalized it to the whole orbit. He did not give adequate empirical reasons for this. Even in his geometrical treatment he had to resort to approximations when he tried to extend the law to the whole orbit. How could he have asserted that it applied to the whole path of the planet in the absence of adequate empirical evidence? Here his philosophical principle of generalizability must have come to his assistance. Since he knew that the law held good for the apsides, this principle enabled him to extend it to the whole orbit.

Kepler's principle of realism undoubtedly played a role in developing his version of the law. For Ptolemy, the relationship was between two empty points, whereas for Kepler it was between two physical bodies. Such a change was only natural for someone who adhered to the principle of realism. For such a person a true scientific law must relate real and physical bodies, not mere abstract entities or empty points.

RELIGIOUS IDEAS

The influence of his religious views was not as conspicuous in this step as in others. Nevertheless, it would be very noticeable in the next one, i.e., in his discovery of the sun as the cause of planetary motion. However, even here the religious element was not altogether absent. I have argued earlier that his belief in the principle of geometrizability had its basis in his religious beliefs, specifically in his idea of God. Insofar as this and similar principles were active in this step, his religious views on which these philosophical and scientific principles were grounded should also be considered active.

ESTABLISHING THE CAUSE OF PLANETARY MOTION

Kepler's Investigations

We have seen that Kepler, unlike the other astronomers of the day, asked the all-important question about the cause of planetary motion. The equant with its Keplerian interpretation gave a first answer to the question, an answer in the right direction. For, as we have already seen, the idea of the equant, coupled with the fact that the maximum and minimum speeds were obtained at perihelion and aphelion respectively, implied that the motion of the planet, at least at the apsides, was directly related to the position of the sun. Not satisfied with this account, Kepler wanted a physical explanation that involved material bodies and real forces and that was formulated in accordance with his principles.

He was concerned with two issues here: the cause of the planetary motion and the nature of the agent (force) responsible for the motion. He answered the first question correctly, but was wrong in his answer to the second. However, since the second question was not essential for the development of his laws, despite this wrong answer he could arrive at correct conclusions concerning the laws.

Already in MC he had wanted to discuss the cause of the equant, i.e., the cause of the truly nonuniform planetary motion. He said that he could not pursue the matter at that time because then it was impossible to decide whether the sun (earth) required an equant.[13] Since that question had been settled, he was now fully prepared to investigate the cause of the nonuniform motion of the planet.

In chapter 33 of AN he argued that the force (*virtus*) responsible for the motion of the planets resided in the body of the sun. He had already established that "the delays [i.e., durations of stay] of a planet on equal portions of the eccentric circle . . . are in the same ratio as the distances of those spaces from the point from which the eccentricity is computed. In simpler words, the farther the planet is from that point which is taken as the center of the world, the less strongly it is moved around that point."[14] This indicated that the center of the universe had a crucial role to play in the origin and distribution of the "force" needed for the motion of the planets. "Therefore," he concluded, "it is necessary that the cause of the diminution in strength reside either in the body of the planet itself, in a motive force placed in it, or at the very assumed center of the universe."[15]

Thus he reduced the possible causes to two: either the planets (their bodies or the "motive force" in them) or the sun. It may be noted that long ago in MC he had made a similar conclusion. After having noted a gradual slowdown of the planets as they moved away from the sun, he wrote: "But if, nevertheless, we wish to make an even more exact approach to the truth, and to hope for any regularity in the ratios, one of the two conclusions must be reached: either the moving souls are weaker the farther they are from the sun or there is a single moving soul in the center of all the spheres, that is in the sun, and it impels each body more strongly in proportion to how near it is."[16] Since he later replaced "souls" with "forces," both the versions in the MC and in the AN were basically the same. Hence Kepler had this idea early in his scientific career.

Kepler now embarked on a long path of philosophical reasoning to establish that of the two alternatives the first had to be ruled out and hence we have to accept the sun as the cause of the planetary motion. He introduced a version of the well-known principle of concomitant variation.[17] In the case under consideration, it has been found that the speed of motion increased or decreased proportionately with the approach or recession from the center of the universe. Hence either the diminution in force was the cause of the planet's recession from the center, or the recession was the cause of the diminution in force, or, again, both effects arose from a common cause. The third possibility he ruled out as impossible and unnecessary. Hence either the diminution in force caused recession, or recession caused the diminution in force. Diminution in force had been found to correlate to distance, so the second alternative said that the slowing down of motion (since recession is accompanied by a slow-down of motion) caused the weakening of force or lengthening of distance. In other words, the second alternative implied that distance was dependent on motion. Kepler said: "Now, it is not in accordance with [the principles of] nature that strength or weakness of motion in longitude be the cause of distance from the center."[18] This is so because distance is logically and ontologically prior to motion, since motion presupposes distance and has to take place within a certain distance. Therefore, the other alternative must be true, i.e., diminution in force caused recession or slow down. Since diminution in force was correlated to distance, he concluded: "Therefore, distance is the cause of the degree of motion. The greater or the less the distance, the greater or shorter the delays [time needed for traversing the path]."[19]

These arguments showed that any attempt to identify the source of planetary motion must focus on distance rather than motion. Since distance is a relative concept—we talk of distance *from* something *to* something—the cause of variation in motion must be found in one of the *relata*. Because the two *relata* were the individual planetary bodies and the body of the sun at the center, there were two possible causes. He first considered the possibility of individual planets being the cause of variation of motion. He rejected this possibility on many counts. One way to account for the increase or decrease in speed of the planets would be by arguing that the body of the planets became lighter or heavier with distance. But this explanation was not acceptable since the heaviness of the body was assumed to be constant. Another possibility would be to argue that the animal force embedded in the planets varied with distance. But he said that such a view would be ridiculous, for if animal force varied with respect to space, i.e., distance, then it should vary also with respect to time, i.e., age. That would lead to instability and chaos in the solar system. Hence the absurdity. There was another difficulty as well. An animal force, according to Kepler, could give only motion, it had no intelligence to guide and control the body. Since Tycho had proved that no solid spheres existed in the sky, the animal force needed an intelligence if it were to move the planets in an orderly manner.[20] Kepler's final reason was as follows: "A round body has no such aids as wings or feet, by the movement of which the soul would be able to carry its body through the aether, as birds do through the air, exerting certain pressure and counter-pressure in that air."[21] Hence he concluded: "Therefore the only remaining hypothesis is this: the cause of the strengthening or weakening [of motion] is in the other endpoint, namely, in that very assumed center of the universe, from which the distances have been computed."[22] Since the center of the system was occupied by no body other than the sun, it followed that "the source of the motive force is situated in the sun, since it has now been placed at the center of the world."[23]

Factors Responsible for the Discovery of the Cause of Motion

SCIENTIFIC OR EMPIRICAL IDEAS

Although the need for philosophical and religious considerations was most conspicuous in the discovery that the sun was the cause of planetary motion, here, too, empirical reasons played a role. We have seen that this idea about the cause of motion occurred to him for the first time in MC, while reflecting on the empirical data of the variation of the period of revolution of the different planets.[24] His study showed that the decrease in the velocity of the planets was not just a simple monotonic function of distance. He noticed that the slowing down of motion was faster than the increase in distance.

Kepler argued that his conclusions could have been reached both by a posteriori and a priori[25] methods: "Undoubtedly, if what I have already proved a posteriori (from observations), by a rather long deduction, if, I say, the same I had attempted to demonstrate a priori (from the dignity and pre-eminence of the sun)...I think I would deserve an equal hearing."[26]

Kepler claimed that he already proved from observations that the sun was the cause of planetary motion. But did he? Could he have done so? Not at all. All that observations could argue for was that the velocity of the planet varied with distance from the central point. This, obviously, was a long way off from accepting the sun as the cause of the variation and, still more, as the cause of all planetary motion. The passage of MC quoted above demonstrates that study based on observational data could not single out the sun as the source of force or cause of motion. That study concluded that the source of force was either in the individual planets or in the central body. For eliminating the first possibility, he needed to carry out a long series of arguments based on philosophical principles. Let us now discuss these principles.

PHILOSOPHICAL IDEAS

Kepler had recourse to the principles of causality, concomitant variation, unity, simplicity, and realism. He had observed that the motion of the planet diminished with distance. From this observation he concluded that the cause of the diminution in strength was either in the body of the planet or in the sun. He could draw this conclusion because he believed in the principle of causality. Again, the principle of concomitant variation played a crucial role in the identification of the sun as the source of force and cause of planetary motion. According to it, if x and y are produced in the same way and at the same time and suffer variation in the same degree, they are interconnected or result from a common cause.

Kepler also used the principle of unity of nature in his arguments. He had to rule out the possibility that the individual planets could be the cause of planetary motion. The arguments he employed in this context presupposed that planetary bodies moved and acted like terrestrial bodies which usually needed legs or wings to move around and which suffered wear and tear with time. Although the belief that the same laws govern both the celestial and the terrestrial domains is quite natural for us, that idea was not common in his time. It was believed that the two worlds were governed by different sets of laws. Kepler, however, believed that the universe is a unified whole and hence the same physical laws govern the whole material universe. It was this belief that enabled him to look at the celestial world in the same way as he did the terrestrial and to develop his argument to rule out the possibility that the planetary bodies were the cause of planetary motion.

Kepler brought up another point often, although not explicitly here: the principle of simplicity. In Tycho's model, if one assigned dynamical causes (which the Dane did not), one had first the sun giving motion to the five planets, then the earth in turn giving motion to the sun. Here two agents were required to account for planetary motion. On the other hand, Kepler's system needed only one agent: the sun was responsible for the motions of all the planets including the earth. In both systems the moon was moved by the earth. Kepler's system showed greater simplicity, and he naturally opted for it.

Kepler's commitment to realism also was decisive in leading him to the conclusion that the sun was the cause of planetary motion. He believed that the cause of motion should be something real and physical. He refused to believe that a vacuous point could be such a cause. According to Ptolemy, the variation of speed took place in reference to the center of the equant, which was but an empty point. Kepler's commitment to realism and search for a physical cause led him to look for a physical body. Thus he could identify the sun as the cause of planetary motion.

RELIGIOUS IDEAS

Although philosophical considerations could rule out the possibility that the individual planets were the cause of motion, they could not have established that the sun and the sun alone had to be the source of force and hence the cause of planetary motion. How could Kepler be sure that the sun alone had the ability to carry out this function? There were difficulties with his arguments to arrive at his conclusion. For instance, in connection with a discussion on the Tychonic system, Kepler wrote in AN: "Therefore, in order that we may not concede that the sun acquires motion from the earth, which would be absurd, we must admit that the sun is immobile. . . ."27 Why should it be so absurd? Kepler did not give adequate philosophical reasons to bring out an absurdity. For instance, the claim that the Tychonic system was less simple could not have made that system absurd. In fact, as we have discussed earlier, the Copernican system also suffered from many serious problems at this time. If Kepler failed to explain the absurdity, he would be begging the question. He seemed to explain this in a letter to Herwart in 1605, where he pointed out that if the sun moved around the earth, it must of necessity be like the other planets and must suffer variation in its velocity. Thus it must move without following a fixed course, since there was none. He continued: "But this is incredible. Furthermore, the sun which is so much higher ranking [solem nobilissimum] than the unimportant earth [terra ignobili] would have to be moved by the earth in the same way as the five other planets are put in motion by the sun. This is completely absurd."28

Hence the reason the earth could not be the mover was that its being the mover would mean that the "unimportant" or "ignoble" earth moved the "so much higher ranking" or "the most noble" sun. The argument here was obviously based on the "higher rank" of the sun. What made the sun to be of a "higher rank?" What philosophical or scientific reasons could be given to prove such a claim? Kepler did not offer any argument, and I do not think he had any strong one to give. In fact, the argument seemed to run counter to his position. For instance, when he wrote the MC, it was believed that the earth controlled the motion of the moon. This showed that the earth was endowed with a force similar to the force that the sun was supposed to have. If the earth could move the moon, why could it not move the sun? The size of the sun could not have been a satisfactory consideration, since Kepler did not have a Newtonian theory of gravitation. In fact, concerning the significance, or rather the lack of it, of size, he had the following remark to make: "But bigness is of no special significance. Otherwise to God a crocodile or an elephant would be more important than humans."[29] As he realized, this point was a strong objection to his position.

Nevertheless, Kepler accepted the sun as the source of force and the cause of motion. Why? I suggest that his idea of the sun was based on his religious beliefs. He was convinced of the nobility of the sun; the nobility could not have come just from size or material. There was a certain divine aura about the "higher rank," as was quite clear when he wrote in the next chapter of AN that "it would seem that there lies hidden in the body of the sun something divine and comparable to our soul. . . ."[30] My view is further confirmed by the fact that in MC he gave the nobility of the sun as one of the main reasons for accepting the sun as the source of force and cause of motion: "Thus just as the source of light is in the sun, and the origin of the circle is at the position of the sun, which is at the center, so in this case the life, the motion, and the soul of the universe are assigned to that same sun so that to the fixed stars belongs rest, to the planets the secondary impulse of motion, but to the sun the primary impulse."[31] Moreover, when he talked of the nobility of the sun, he made sure that the reader would not miss the divine symbolism: "Consequently the sun has a far better claim to such epithets as heart of the universe, king, emperor of the stars, visible God and so on. But the nobility of this theme demands a far different time and place. . . ."[32] The sun for him was the primary mover, in fact the unmoved mover. We have seen earlier that for him the unmoved mover was the visible parallel of the invisible uncreated creator, i.e., God the Father. Thus his claim that the sun had a "higher rank" was closely associated with the claim that the sun in a special way represented God.

Kepler concluded that the sun moved the planets by means of a *species* emanating from the solar body and spreading like light in the space all around the sun. He could arrive at this conclusion because he gave a Trinitarian interpretation to this operation: the Father represented the source at the center; coming from this source were the divine rays or emanation, which filled the intervening space; this emanation-filled space itself stood for the Holy Spirit. Kepler believed that the sun's emanation had the power to move the planets, because just as the Holy Spirit is all powerful, the *species* emanating from the sun and filling the space was also powerful. Therefore his religious belief in the parallelism between the spiritual, heavenly world and the physical, material world played an active role in identifying the sun as the source of force and cause of planetary motion.

STUDYING THE ACTUAL PATH OF THE PLANET WITH THE HELP OF THE DISTANCE LAW AND THE CAUSE OF PLANETARY MOTION

Kepler's ambition was to find out the exact path of the planet around the central body. He now had at least three important tools in his hands: a law governing the motion of the planet, a clear idea about the source supplying the force needed for the motion, and the idea about the correct point of reference with respect to which measurements and calculations could be made. Equipped with these newly acquired tools, he embarked on his project. His basic procedure consisted in first assuming a shape for the orbit and presupposing the inverse distance rule, and then seeing if he could obtain planetary positions that agreed with the vicarious hypothesis. In order to carry out his study he divided the eccentric orbit into 360 parts of one degree each. He then determined the length of each of the small parts and added them together.[33] He thus performed the long and arduous calculations to get the equation or anomalistic difference of the planetary motion. This method of equation basically determined by what angle the planet's position differed from its mean position. He found this procedure "mechanical and laborious. Nor is it possible to compute the equation of any particular degree, without taking into account all the other degees. . . ."[34] He looked for another method which would simplify calculation, and his move towards simplification was a stroke of genius, for it led him directly to his second law.

DISCOVERING THE AREA LAW
AS A METHOD OF SIMPLIFICATION

Kepler described the method of simplification: "When I realized that there is an infinity of points on the eccentric, and therefore an infinity of distances, it occurred to me that the surface of the eccentric comprised them all, for I remembered that Archemedes of old divided the circle into an infinity of triangles, when he sought the ratio of the circumference to the diameter."[35] Thus, instead of dividing the orbit into 360 parts, he cut the total area of the orbit into as many sectors by drawing lines starting at the point from which the eccentricity was measured. These areas could be calculated without difficulty, and the procedure or method could give the equation of a planet in an eccentric circular orbit. By using the idea that the area of a sector of the orbit provided a measure of the time needed for the planet to traverse the corresponding segment of the orbit, he could simplify the laborious process. This was the beginning of Kepler's second law. Later, after he had discovered the law of the ellipse, he would give a formal, geometrical proof for this law.[36]

Koyré has pointed out a problem here. Strictly speaking, the summing of distances should involve summing an infinite number of distances because an orbit has an infinite number of points. Kepler had a real phobia for infinities; he considered the infinite to be indefinite and unintelligible. In his method, the problem was really serious. No matter what the length of the segments in question, the sum involved would be infinite, since every segment would have an infinite number of points and hence no ratio could be established among them. Koyré remarks that Kepler "was not aware of the fact,"[37] a puzzling remark because Kepler certainly was aware that a method of this type, in principle, involved an infinite number. If he had forgotten this point, his relentless correspondent Fabricius was there to remind and attack him on it. Kepler was not perturbed by the infinity problem because he believed that the discrepancy involved was negligible. So he wrote to Fabricius in 1603: "You go now and compute either the infinite points of its [Mars's] route or the arcs cut into the very smallest sections, say into ten minutes [scrupula], and you will find the same thing I found with my vicarious hypothesis, with scarcely half a minute's difference."[38] Hence he felt justified about his division of the orbit into a finite number of parts. He believed that by this method one could get practically the same results as one would, if one were able to divide the orbit into an infinite number of parts. The problem of replacing area by length is equivalent to the summing of infinitesimals, and, of course, Kepler was not alone in the confusion regarding the incipient techniques, which would become the integral calculus.

Kepler himself pointed out that there were problems with the method. He admitted that he was following the example of Archimedes. However, the infinitely many triangles Archimedes had constructed by dividing a circle were right angled with respect to the circumference and had their apices at the center of the circle. But in the case of Kepler's orbit the apices of the triangles were at an off-center point, i.e., the position of the sun. Hence the straight lines drawn from the sun to the circumference cut it obliquely in all points except at the aphelion and perihelion. Thus the triangles produced were not right angled and could not strictly be compared to those produced by the Greek. Kepler expressed the error involved here in another way. Assuming that the radius was equal to 100,000, if we added all the distances from the center of the circle to termini of the one degree segments of the circumference, we should get 36,000,000 exactly. He found that when the distances were taken from the sun and added together, he got a sum greater than 36,000,000. This showed that the two sums were not a measure of the same surface.

Kepler argued that this error "is of no great importance,"[39] not only because the difference in areas involved here was not very great, but even more because two different errors were involved and they, "as though by a miracle, cancel each other most exactly."[40] The two errors he had in mind were these: First, the assumption of the circular orbit. At this point of his research the circularity principle was still held. Since the path of the planet was not circular, the assumption brought in an error. Second, the method made use of a surface which did not exactly measure the distances from the sun to all the points of the path.[41] Thus the area here failed to measure exactly the sum of the distances from the sun. This introduced another error. Kepler got a lucky break because these two problems erred in the opposite direction and by the same amount, thereby canceling each other.

FACTORS RESPONSIBLE FOR STUDYING THE PATH OF THE PLANET AND DISCOVERING THE AREA LAW

Since the steps discussed above—studying the actual path of the planet and discovering the area law—are naturally and intimately interrelated, the factors responsible for them can be discussed together.

Scientific or Empirical Ideas

Empirical considerations certainly functioned in the last two parts involved in Kepler's discovery of the second law. He checked with observed data the values he obtained through mathematical study. He accepted the area rule and what he considered to be the actual path of the planet because he found that the discrepancy between the calculated values and the observed values was not large.

As always, he looked for explanation based on tangible, real explanatory factors. The area law seemed to have provided him with such an explanation. He often used the word "distance" in a concrete sense referring to an actual line segment or an actual "ray" of the emanation.[42] Hence the area under consideration was not an abstract space but was occupied or filled with the *species*. This idea he made clear in his letter to Fabricius: "By collecting all the distances which are infinite in number, one gets the sum of the force spent at a certain time and therefore also of the path traversed around the center of the eccentric (and also around the sun)."[43] The area under consideration was filled with the nonmaterial, but matter-directed and real force. (This idea is quite similar to the electromagnetic field of later centuries). Obviously, here one can see the influence of his insistence on the physical and the tangible. Not something abstract, the space here served as a link between the moving sun and the moved planetary body.

Philosophical Ideas

Geometry played the most crucial nonempirical role in the last two steps of the discovery of the second law. As we know, geometry was deeply rooted in Kepler's belief that the universe is geometrical and hence geometry held the key to understanding the universe. This principle of the geometrical nature of the universe in turn was based on his religious belief in the geometer God. It follows, therefore, that his use of geometry had both a philosophical and religious dimension.

One may look at Kepler's adoption of the area law as a mere labor-saving short cut. However, the mere pragmatic consideration of economy of labor alone could not have induced him to accept this law. Indeed, hard labor was painful for him, as it is for everybody else. But fear of hard work never deterred him from pursuing his goal. As he told Longomontanus, if he were frightened of hard work, he could have ignored the error of eight minutes and that would have reduced his work to a fraction—in fact, to a mere third.

The adoption of the area law had another problem as well, the apparent violation of Kepler's principle of precision; as he himself admitted, the whole procedure involved approximation and even two errors, which fortunately canceled each other. Yet he accepted the area law. I think that a probable explanation for this strange behavior can be found in Kepler's deep commitent to the principles of simplicity and economy, principles extremely important for him. Since nature is simple, the laws of nature must also be simple. A move that can simplify calculations should be true, provided that it agrees with observations better than all its rivals and that it is in agreement with the overall worldview.

A remarkable point we see here is Kepler's willingness to compromise some of his principles in favor of others. Although he took the principles seriously, he was not so fanatically attached to them as to remain immovable from them. Here we see him willing to compromise some aspects of the principle of precision for the sake of the principle of simplicity. Later on, when he adopted the elliptical orbit in place of the circular, we see him willing to compromise the principle of simplicity. Openness to alternatives and inner freedom to adjust and adapt to new findings were two of the secrets of his success.

Religious Ideas

Perhaps the most important reason he could accept the compromise and adopt the simplification and hence the area law was that the law, according to him, agreed marvelously well with his theory of the sun as the source of force and cause of motion, as the cause of planetary motion (see above). He considered this agreement crucially important. Since the sun occupied a dominant place in his system, a theory that assigned a preeminent role to the sun in effecting planetary motion was closest to his heart.

The considerations above show that in the discovery of the second law all the three factors—empirical, philosophical, and religious—were active. The empirical alone could not have accomplished this; for instance, it could not have identified the sun as the source of force and cause of motion. The philosophical could also help him weed out the different possibilities in his search for the cause of planetary motion. His philosophical view further assured him of the validity and need for the use of geometry in his investigations. In a unique way, religious beliefs enabled him to single out the sun as the "unmoved mover" of the planetary system.

9

THE DISCOVERY OF THE FIRST LAW

FROM THE CIRCLE TO THE OVAL

The Significance of the Discovery

Although Kepler introduced several revolutionary ideas into astronomy and physics and thus in several ways transformed our view of the world, none of his contributions was so radical as his rejection of the age-old idea of the circular orbits of the planets. The circularity principle, the maxim that all planetary bodies, despite their appearance to the contrary, in reality moved in circular orbits, was so deeply entrenched that no one had dared to challenge it. Copernicus's heliocentrism was acclaimed as revolutionary, but it, unlike Kepler's challenge to circularity, offered nothing new or unheard of; down through the centuries many eminent thinkers and astronomers had proposed a heliocentric view. With regard to the circularity principle, however, as Hanson puts it, "In 2000 years of technical, computational astronomy it had never been questioned."[1] Not even a daring revolutionary like Galileo was prepared to challenge the principle. Ptolemy's introduction of the equant and consequent acceptance of the nonuniform motion of planets around the center of their orbits did great violence to the "sacred" principle of uniform planetary motion, but even this innovation left the circularity principle intact. In fact, as Brackenridge points out, Ptolemy's equant only separated the uniformity of movement from the circularity of movement.[2] Kepler's rejection of such a deep-rooted principle, universally regarded as inviolable, was no easy task for him. He tenaciously held onto it until very late in his career and parted with it only after several different kinds of evidence convinced him of the untenability of the principle. This chapter is a detailed study of the reasons that led him to make such a momentous decision. I shall argue that he could make this move, which neither his predecessors nor his contemporaries had dared to make, because in his thoughts and works he allowed the interplay of different sources of knowledge.

The variety of reasons—observational (scientific), philosophical, and religious (theological)—on which belief in the circularity principle was based could explain why it continued to enjoy universal acceptance for so many centuries. Several arguments on the basis of observation seemed to show that the heavenly bodies described a circular path. The diurnal rotation of the heavens was taken as firm evidence that the bodies in the heavens followed a circular path. Again, observations showed that constellations like Draco, Cepheus, and Cassiopeia executed circular motion around the poles.[3] Stars which were not circumpolar regularly rose and set at the same place each night. Such periodicity and regularity could presumably be explained only in terms of circular motion. From the sixth century B.C. onwards it was known that the earth is spherical, because ships approaching from the far sea first expose their mast.[4] Like the earth, the heavenly bodies were also considered to be spherical, and it was believed that perfect spherical bodies performed perfectly circular motion.

The philosophical reasons for the circularity principle were most emphatically propounded by Aristotle. According to him, circular motion was the primary and most perfect motion: "Circular motion is necessarily primary . . . the circle is a perfect thing. . . ."[5] By perfect motion he meant a continuous motion, one without termini since a motion which began and ended at discrete points would be incomplete and imperfect. "Circular motion since it is eternal and perfectly continuous, lacks termini. It is never motion towards something. Only incomplete, imperfect things move towards what they lack."[6]

The nature of the element constituting the heavens also argued for this principle. The heavenly world was made up of a fifth element, the characteristic natural motion of which was circular. Obviously, this principle was nurtured by the overwhelming authority and often-repeated arguments of Aristotle.

There were other reasons as well for believing in circularity. The circle is a perfectly symmetric figure, rendering it most pleasing to the order- and harmony-loving philosophers. Furthermore, it was believed that all useful geometrical figures could be produced by compass and ruler, which seemed to assign fundamental importance to straight line and circle.

Religious reasons also dispelled any doubt concerning the universal validity of this principle in heavenly motion. The heavens were the abode of the divine beings. Nothing imperfect was admissible there. Since the circular motion alone was the perfect motion, that alone could be allowed there. Aristotle presented yet another argument: "The movement of that which is divine must be eternal. But such is the heaven, viz., a divine body, and for that reason to it is given the circular body whose nature it is to move always in a circle."[7]

Of course in the old geocentric system the marked deviation from circularity in the planetary path actually observed from the earth never went unnoticed. However, astronomers did not allow these deviations to threaten this inviolable principle. It was stipulated that these irregularities should be treated as only apparent, to be accounted for in terms of regular circular motions. Hence epicycles were introduced into the arsenal of astronomical explanation.

To us living in the twentieth century and enjoying the benefit of hindsight, the arguments in favor of circularity look very unconvincing, to say the least. However, that was not the case in the days of Kepler, when Aristotle's authority still exerted a powerful influence, especially because no viable alternative physics was available at the time.

That the influence and importance of this principle remained unabated is evident from the total adherence to it of both Copernicus and Tycho. The former's idea and arguments for the principle were exactly the same as Aristotle's. In fact, in DR Copernicus almost verbatim repeated the Philosopher's arguments. For instance, the Canon wrote in book 1, ch. 4: "We now note that the motion of heavenly bodies is circular. Rotation is natural to a sphere and by that very act is its shape expressed. For here we deal with the simplest kind of body, wherein neither beginning nor end may be discerned nor, if it rotates ever in the same place, may the one be distinguished from the other."[8] According to him also, rotation was part of the very nature of a sphere: a spherical body rotates simply because it is spherical. Like the astronomers before him, Copernicus also admitted that the observed path of planets was irregular and, once more like them, refused to accept this irregularity as a counterexample. Hence he continued:

> Nevertheless, despite these irregularities, we must conclude that the motions of these bodies are ever circular and compounded of circles. For the irregularities themselves are subject to definite laws and recur at stated times, and this could not happen if the motions were not circular, for a circle alone can thus restore the place of a body as it was. . . . Now therein it must be that diverse motions are conjoined, since a simple celestial body cannot move irregularly in a single circle. For such irregularity must come of unevenness either in the moving force (whether inherent or acquired) or in the form of the revolving body. Both these alike the mind abhors regarding the most perfectly disposed bodies.[9]

Tycho also held similar views, despite his hitherto unrivaled ingenuity as an accurate observer of heavenly bodies, and despite the fact that in certain aspects he rejected the traditional Aristotelian position, e.g., the solid sphere theory. In a letter to Kepler, Tycho wrote: "It is necessary that the orbits of planets be composed completely of circular motions, otherwise they would not return perpetually and uniformly on their courses, and their perennity would be disturbed."[10] According to him, a noncircular orbit was simply unthinkable: "A rational mind would recoil with horror from such a supposition [of noncircularity]."[11] Hence the irregular motions would have to be explained in terms of regular, circular motions. Indeed, the real motion was circular and any irregularity observed was only accidental. "If the circular motion in the heavens by the manner of their arrangement sometimes seems to produce—to anyone who had the vain curiosity to note such oddities—diverse angular figures mostly of oblong shape, then that can only be by accident. . . ."[12] Notice that, despite all the importance he gave to accurate observations, he discouraged any serious study of the deviation from circularity, since according to him, only those with "vain curiosity" would indulge in any serious consideration of such "oddities." His commitment to the circularity principle seemed to have been so deep that it became impervious to any objections, even on the basis of accurate observations. In his view circular motion was best suited for scientific study. The less simple and regular the planetary courses were, the less suitable they were for scientific study and computation.

This was the tradition Kepler inherited and this was the intellectual milieu in which he lived and worked. In fact, in the beginning he himself subscribed to such a view. In the MC he talked of planets: "But if it is assumed that the planets required motion, it follows that they had to receive round orbits in order to acquire it."[13] Right at the beginning of the first chapter of AN he affirmed the principle: "We have the testimony for ages that the motions of planets are orbicular. Reason derived from experience unhesitatingly takes it for granted that the gyrations of planets are perfect circles. For among all the figures and bodies in the heavens, the circle is regarded as the most perfect."[14] In some ways he was a more faithful adherent of the circularity principle than most of his contemporaries. Koyré points out that, although most astronomers long ago had ceased to insist that the effective orbits of the planets must be circular, Kepler continued to demand it.[15] Hence it was a millennia-long, well-entrenched tradition, a tradition to which he himself had adhered most faithfully for years, that he broke away from. Let us now discuss the reasons that motivated him to make the revolutionary departure from tradition.

Kepler's Investigation

While continuing his investigation on the path of Mars, Kepler noticed that the observational values did not agree with the theoretical or calculated ones. His procedure was as follows: he presupposed the vicarious hypothesis and his solar theory (i.e., the theory of the earth developed previously). The vicarious hypothesis should give him the heliocentric longitude of Mars and the latter the positions of the earth. In this situation three observations of Mars when not in opposition to the sun should be sufficient to determine three positions of the planet, through which only one circle could be traced. Thus assuming that the path of Mars was a perfect circle, he proceeded to determine its exact position, the eccentricity, and the position of the line of apsides. Using three accurate observations he determined the path. However, he noticed that if he replaced these observations partially or completely with others, he could not get the same results. The results obtained differed significantly. For instance, in the first set of observations the eccentricity was 9768 whereas in the second it was only 9264 (the radius of the orbit being taken as 100,000). Now if the orbit was a circle, then three accurate observations must have been sufficient to determine the orbit completely and definitely. The result gave rise to a strong suspicion that something was wrong with the assumption of a circular path.

The negative results of the method forced him to make a fresh move. He wanted to apply the method of areas in this investigation, just as he had done in studying the earth. At first, he decided to find out whether the bisection of eccentricity was applicable for Mars, just as it was for the earth's orbit. He decided to determine the aphelion and perihelion distances on the basis of observations alone, without assuming any specific shape for the orbit. From Tycho's treasure he was able to pick out nine observations for the determination of the aphelion and three for that of the perihelion. By means of these observational data he was able to establish that the distance of the aphelion of Mars from the sun was 166780 units and that of the perihelion 138500 units. He first assumed this value and then by trial and error established that the assumption was correct. Hence the semidiameter would be (166780 + 138500)/2 = 152640 and the eccentricity would be 152640 − 138500 = 14140.[16] If he had assumed the radius of the orbit to be 100,000, then the eccentricity would be 9264. On the other hand, the eccentricity of the equant given by the vicarious hypothesis was 18564, half of which would be 9282. Since the difference between these two results was only 18 units, he could conclude that the bisection of eccentricity held good in the case of Mars as well.

These studies enabled him to believe that Mars was governed by the same laws as the earth. Furthermore, his physical interpretation of the bisection of the eccentricity assured him that the sun was the cause of motion and source of power and that the speed of motion varied inversely with the distance from the sun. All these considerations paved the way for the use of the new physical method, which utilized the idea that the speed of motion of the planet was a measure of the area swept by the radius vector, in the investigation of the equation of motion of Mars, just as he had done in the case of the earth.

His application of the new method yielded results which fitted well for the case of mean longitudes. For instance, for the eccentric anomaly equal to 90 degrees the difference was only 24 seconds. However, the agreement with the vicarious hypothesis was not so good for other points of the path. For eccentric anomaly equal to 45 degrees, the discrepancy was 8 minutes and 21 seconds, and for 135 degrees it was 8 minutes and 1 second.[17]

Since such errors were not acceptable to Kepler, he began looking for possible causes of the errors. Naturally, the first candidate was his own new method. Could the substitution of areas for "sums of distances," which was not fully justified, be the culprit? He found that this was impossible. The error arising from this approximation amounted to no more than 7000 parts of the semicircular area, if the total area of the semicircle was taken as 18,000,000. When translated into angular units it would be 4 minutes and 12 seconds, a mere half of the discrepancy in question. Furthermore, this discrepancy could not be accumulated at any particular region, but must be distributed over the entire semicircle. Hence the actual error at the points under consideration must be even smaller. Besides, if the error were due to the new method, then the time taken at the intermediate or mean position (90 degrees) would be shortened. However, it was found that the error made it too long.

Could the error have come because of his rejection of the double epicycles of Copernicus and Tycho? Kepler ruled out this possibility because the double epicycle model gave rise to a path which ran out beyond the circle at 90 degrees of eccentric anomaly by 246 parts of 100,000.[18] Since the use of the method of areas would shorten the distances at the mean positions, the effect of this method should have been to make it run out beyond the circle even faster. The acceptance of the Copernican double epicycle would only have increased the error, rather than decreased it, so the source of the error could not be attributed to the rejection of that model. As Kepler himself put it, "the Copernican orbit does not curve towards the center, as we require here, but, on the contrary, runs further away from the center by 246 parts [of the radius]. This would only increase the error still further, since here we assume that the delays [i. e., times spent or required for traversing specific arcs of the orbit] are proportional to the distances."[19]

Another challenge to the circular orbit came from Kepler's consideration of the distances obtained by triangulation.[20] In figure 7, S is the sun, O is the center of the circular orbit, and DSOC the line of apsides. Let E, F, G be three positions of the planet on a perfectly circular orbit. Kepler found that the distances such as SE, SF, and SG, deduced from observational data, did not agree with those obtained by calculation on the basis of a precisely circular orbit. In fact he found that the calculated value of SF exceeded the observed value by 350 units. Similarly, in the cases of SG and SE the discrepancies were 783 and 789 units respectively.[21] This showed that the path of the planet deviated from a perfect circle and the deviation was least at F, a point close to the aphelion, whereas at G and E, points close to the midpoint between the apsides, the deviation increased considerably.

Kepler argued that these discrepancies could not have been due to observational errors. They were too large to be ignored as errors of observation. According to him, "if anyone should wish to ascribe these differences to errors of observations, he would have to betray the fact that he had neither paid attention to nor understood the force of our proofs; and he would have to impute me with the most henious fraud of distorting most grossly Brahe's observations."[22] All these considerations led Kepler to conclude: "Therefore it is plainly clear that the planet's orbit is not a circle; but [starting at aphelion] it curves inwards little by little and then [returns] to the amplitude of the circle at perigee [perihelion]. An orbit of this kind is called an oval. . . . The orbit of the planet is not a circle, but an oval."[23]

The next step in Kepler's investigation was the search for the causes for this deviation from the circular path. Thus he began to study the mechanism by which such an oval motion could be explained physically.

Factors Responsible for This Discovery

SCIENTIFIC OR EMPIRICAL IDEAS

From what has been said above one may conclude that it was purely on the basis of observational data that he rejected the circular orbit and accepted the oval. In fact, some writers have drawn

FIGURE 7

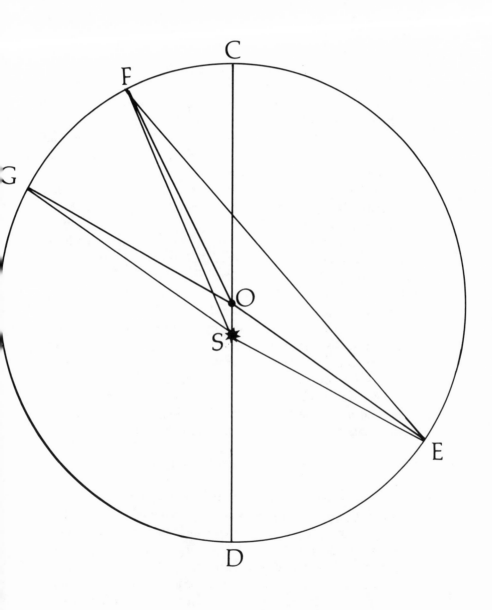

such a conclusion.[24] However, such a conclusion is inaccurate, both historically and logically—historically, because Kepler knew that the observational data available to him were inadequate to draw his conclusion; logically, because strict logic need not have led him to the oval on the basis of observations alone.

Kepler's contemporary Fabricius, himself a well-respected astronomer of the day, presented several arguments based on observations to refute the oval. In his letter to Kepler on Oct. 27, 1604, he subjected the oval hypothesis to observational test and found that observations did not support the German's innovations.[25] It is remarkable that in his reply Kepler did not say that Fabricius was mistaken, that he had absolutely reliable observations to establish his oval. Let us listen to his own words: "You [Fabricius] argue faultily [*vitiose*], 'Kepler's oval falls too short, hence one cannot simply assume it is oval.' I argue equally faultily [*vitiose*], 'it is some kind of an oval, hence it will be the one which uniform epicyclic motion demands.' A perfect circle exceeds the dimensions of the annual orbit (taken to be 100,000) by about 800 or 900. My oval falls short by about 400."[26] From this he concluded: "The truth is in the middle, but closer to my oval."[27] Clearly, according to Kepler, from the observational point of view both the circular and the oval hypotheses erred, but the oval erred much less. Observations could help him only to choose the less inaccurate hypothesis. In other words, all that observations could establish was that the oval hypothesis was better than the traditional one and hence was highly probable. Since being better or being highly probable need not render the hypothesis correct, observational data were inadequate to establish the oval orbit definitively.

PHILOSOPHICAL IDEAS

A close look at Kepler's research and reasoning at this stage reveals that philosophical considerations supplied some additional arguments that he needed to draw his conclusions. Specifically, Kepler came to realize that no purely natural means existed by which a perfectly circular orbit could be produced.

Kepler had already convinced himself that the sun gave out an immaterial but real *species* distributed circularly, which provided the pushing force responsible for planetary motion. However, there was a serious problem: on the one hand, the solar "whirlwind" was distributed circularly and uniformly and so its effect must be uniform; on the other hand, it was noticed that the planets showed variation in their velocity around the sun and in their distance from the sun. Since the solar force was neither attractive nor repulsive, there must have been perfectly uniform motion, if it depended only on the sun's *species*. Also, the distance from the sun should have remained constant and invariable. This state of affairs prompted Kepler to ask the question: "If there are no solid spheres, as Brahe has shown, how come that the planet moves nearer to, or away from the sun? Can it be the case that this result also is produced by the sun?"[28] He admitted that to a certain measure the sun was responsible. But more than the sun's influence was involved here.

He proceeded to give an argument that would both preserve the principal position of the sun as the cause of motion and would allow the introduction of some other agent to account for the variation of planetary motion: "The examples of natural things and the relationship between the celestial and the terrestrial declare that the workings of simple bodies are the simpler, the more universal they are. Variations, if there are any (as in the case of planetary motion, where there is variation of distance from the sun, or eccentricity) arise from concurrent external causes."[29] He seems to say that in the case of simple bodies, the more universal (*communior*) the applicability of a theory, the simpler it is; the more restricted, the more complex. Thus this principle not only implies a direct correlation between universalizability (universal applicability) and simplicity, but even more seems in fact to advocate that simplicity is a basic characteristic of the universally applicable principles. This certainly is intimately related to his favorite principle, "Nature is simple," which says that the universe must be simple; the principle Kepler used here says that what is universal must be simple. Laws of nature which are universally applicable are considered to be simple.

Since the solar emanation is most universal, it must be most simple. On the other hand, deviations in the planetary motion are restricted and limited, and hence must not be simple. Hence the universal influence of the sun alone could not be the cause of planetary approach toward and recession from the sun. This variation must be attributed to some other agents. To illustrate his point, he gave the example of a ferry boat crossing the river with the help of a pulley moving on a cord stretched across the river, the boat being connected to the pulley.

In this setup, the ferryman can cross the river by suitably arranging the rudder. Here the main force is supplied by the current of the river, which is continuous and uniform. The cord, pulley, rudder, and the ferryman supply the additional force needed to cross the river. These considerations led him to his conclusion: "Therefore it is necessary that the planets themselves, like small boats, have motive forces of their own, as if they had passengers or ferrymen, by whose forethought they accomplish not only the movement towards, or away from, the sun, but also. . . the declinations of latitudes."[30] Thus Kepler argued that to account for the planetary motion adequately, one had to invest the planets with individual forces of their own.

Having established this point to his satisfaction, he wanted to investigate how the two forces, the one from the sun and the other from the individual planets, acted together to produce circular planetary motion. It was well known that an eccentric circular motion could be produced with the help of an epicycle, provided the center of the epicycle revolved on the deferent and the planet itself moved on the circumference of the epicycle in the opposite direction, both the center of the epicycle and the planet always covering similar arcs. This was the possibility he investigated next: "If it were possible for the planet to describe a perfect epicycle by its own inherent force, while at the same time its orbit remained a perfect circle, then it would be necessary to think that similar arcs are described in the same time on both the epicycle and the eccentric."[31] This would be so if circular motion of the planet was to be preserved. Hence we have to ascribe the same velocity to both the planet and to the center of the epicycle, meaning that these two motions would accelerate and retard simultaneously.

The above mechanism of epicycle-on-the-deferent could bring about perfectly eccentric circular motion. But according to Kepler, this gave rise to grave absurdities. He explained one of them as follows: The center of the epicycle in its motion along the deferent and the planet in its motion around the center of the eccentric would have to have the same angular velocity. Hence the two motions would accelerate and retard simultaneously. "Since the acceleration and retardation are due to the greater, or less, distance of the body of the planet from the sun, the center of the epicycle, always remaining at the same distance, is made to move with faster or slower speed according as the planet is farther away from, or nearer to, the sun."[32]

Therefore, on the one hand, the center of the epicycle was at a constant distance and hence should move with a constant velocity, since the motion depended only on the distance from the sun; on the other hand, it had to move with a variable velocity. Absurdity.

Kepler pointed out that the matter could be simplified if one directed attention to the diameter (or radius) of the epicycle rather than to the epicycle. In this case the planetary intelligence had only to make sure that the diameter (or radius) remained parallel to itself and kept the planet at a fixed distance from the center of the eccentric and from that of the epicycle. However, this also would lead to an absurdity because it implied that the planetary intelligence determined the distance between itself and the empty central point, "where no particular body exists as an indicator."[33] He admitted that he was not denying the possibility of imagining such a center. But he believed that it was impossible for a real body to execute a perfectly circular motion about a purely imaginary point.

All these considerations led Kepler to the following conclusion: "The epicyclic motion required to realize an eccentric and perfectly circular orbit cannot be produced by purely natural means."[34] He clearly admitted that, if what we were looking for was a geometrical explanation, as the ancients did, an adequate answer could be obtained; but if we looked for a physical explanation, then there was no satisfactory means by which a circular motion could be produced.

Why did Kepler see these absurdities? Koyré points out that the situation caused no awareness of absurdity to either Ptolemy or Copernicus.[35] According to Copernicus, there was no absurdity in an epicycle and its center moving with equal angular velocities, provided they were constant. Hence the epicycle-on-the-deferent mechanism need not give rise to any problem. Ptolemy would have no difficulty with a nonuniform motion of the center of the epicycle; in fact, this was precisely what the equant revealed.

I think that Kepler's worldview and his idea of an adequate explanation were at the basis of his perception of the absurdity. Kepler's belief in the principle of realism demanded that an adequate explanation of phenomena be in terms of real physical forces: a mere geometrical explanation could not satisfy him. Indeed, he emphasized geometry, but only as a necessary condition, not a sufficient one. "Ubi materia, ibi geometria," he had declared. Where matter is present, there geometry is also present—a physical geometry, not just pure, abstract geometry. He wanted only those geometrical figures which could be produced naturally, by compass and ruler. Commenting on the epicycle-on-the-deferent model, he wrote: "In this way, we should, doubtless, come close to the geometric concepts [*suppositionibus*] of the ancients, but we would stand very far aside from physical considerations."[36] According to him, the path of the planet had to be the real orbit, not some purely geometrical one. Being in agreement with certain laws or concepts of geometry was not enough for the path; it also had to be consonant with the principles of realism.

One way to get out of the absurdities would have been to argue that the heavenly bodies followed a different set of laws, the argument often resorted to by the Aristotelians. However, Kepler believed in the principles of unity and uniformity of nature. Hence in his view the laws governing heavenly bodies were not different from those governing the terrestrial bodies. This implied that the requirement of physical, dynamic explanation was equally applicable to the planetary phenomena. Hence it was absurd that the motive force should act on an empty, imaginary center of epicycle, a point bereft of any physical reality. This absurdity would become worse if it should cause the vacuous, imaginary point to move with varying velocity. Still worse would be to have to suppose that the motive force acting on the planetary body could give rise to a steady motion of the imaginary point.

Having discovered that the consideration of the epicyclic motion could not be of much help, Kepler turned to the possibility of libration of the planet on its diameter or radius vector. This method, indeed, would require planetary intelligences. But since he was not opposed to that idea,[37] this requirement posed no serious difficulty. However, there were other serious problems. First of all, for this method to work one had to assume that the planetary intelligence had to memorize a table of equivalents similar to the Alphonsine or Prutenic tables. Although this would be almost impossible in practice, at least in principle it could be done. The major problem was that libration required a translatory motion, a motion from one place to another; but the animal force inherent in the individual planetary body could not give such a translatory motion.[38] Hence Kepler was once more convinced that there was no natural way of producing circular planetary motion.

The arguments from empirical and philosophical sources could not have established the oval orbit conclusively. Kepler himself admitted this fact because when he was attacked by Longomontanus and Fabricius, he did not hesitate to acknowledge that problems remained. He compared his project to the Herculean task of cleaning up the Augean stables. He admitted that, although he had cleaned up most of the mess, there still remained one "cart-full of dung." He replied to Longomontanus: "Must one be punished for leaving one cart of dung, although the rest of the Augean stables have been cleansed? In this sense, why do you reject my oval or the one cart of dung, when you have put up with the spirals which are the entire stable, if my oval is only one cart?"[39]

RELIGIOUS IDEAS

If arguments based on empirical and philosophical ideas, either individually or in combination, were not powerful enough to persuade Kepler to make this move from the circle to the oval, how could he have abandoned the circle? The gravity of the situation becomes even more puzzling when we realize that this imperfect model was adopted to replace the millennia-old tradition of the circular orbit. How could he have accepted this model for such a momentous purpose? Could it have been a strange streak of irrationality or mysticism? I do not think so, because at this stage of the AN, he was perhaps at the peak of his rationality; more than ever at this stage he emphasized the need for rationality. For instance, so often when pressed by the burden of the long and complicated calculations of the oval orbit, he would mournfully exclaim: "If only the path were an ellipse!" An elliptical path would have simplified his work. But he would not take the easy way out, because at this stage his reasons pointed to the oval path and he did not find adequate reason to believe that it could be elliptical.[40] Without adequate reason he would not accept something even if it was extremely beneficial. We hardly find any mystical streak at this stage of his work. At other stages of his life and work, he would often suddenly interrupt his strictly scientific line of thought and burst into sentiments of praise and worship of the Deity, but such outbursts were extremely rare in this part of the AN. Hence his acceptance of the oval wholeheartedly—again and again he affirmed that the path of Mars was an oval—despite the unanswered or underanswered objections, was not based on any irrational or mystic wave.[41]

A close and careful study of Kepler's thought reveals that his religious views supplied him with the additional arguments necessary for reaching his conclusion. More specifically, his belief that the human soul or mind is the "incorporeal image of God," whereas the material universe is "the corporeal image of God," played a significant role in convincing him of the acceptability of his conclusion.

As I have already argued, the empirical evidence established that oval orbit was better, i.e., far more acceptable, than the traditional circular one. His philosophical convictions led him to conclude that it was impossible to produce a perfectly circular motion by purely natural means. The basic form of his argument was this: Not only was the oval superior to the circular, but the latter could not even be produced, and hence the oval must be accepted. The argument presupposes that the oval and the circular were the only two contestants. This is debatable, but since according to Kepler, at this stage these were the only two viable and serious options, we may concede the premise. But how could he establish that there was no natural means to produce the circular orbit? Because the consequence of the conclusion was so momentous, since it would lead to the rejection of the principle of circularity, he had to be absolutely sure of his claim that there was no natural means for producing the circular orbit. How could he know this? How could he be sure that he had investigated and exhausted all possibilities?

Maybe there existed some mysterious method unknown to him. This possibility did not bother him because he believed that the human mind was capable of knowing the means of production of circular path if it really existed. The human soul, the incorporeal image of God, shared in the knowledge of God. God created the universe and for this creation made use of the archetypes eternally existing in God's own mind. The human mind or soul shared in these archetypes and so should be able to know the operations by means of these archetypes. Indeed, Kepler did not want to say that humans were omniscient like God; such heretical views were far from him. However, he did believe that God used geometrical archetypes and, where these were involved, humans had access to God's knowledge.

Kepler's emphasis on the aspect of "natural" production was also relevant here. Natural was opposed to purely abstract or hypothetical: natural for Kepler meant real. His stressing "natural" was another way of expressing his commitment to realism, his belief that the universe is real and the phenomena in it are brought about by real means. This conviction was fundamental in his thought because for him the universe was the corporeal image of God. Since God is supremely real, the world should also be real.

Once these considerations are taken into account, Kepler's conclusion will not seem odd or unwarranted. For if the circular path is produced by real natural means and we can know them, provided they exist, then our failure to find them after repeated efforts should lead us to infer the nonexistence of such a path. Thus in his acceptance of the oval, his religious views also made a contribution, though in a less striking way than in other discoveries. Hence Kepler's most original move, his break with the circular orbit and his acceptance of the oval, resulted from the combined influence of empirical, philosophical, and religious ideas.

FROM THE OVAL TO THE ELLIPSE

Although Kepler was correct in breaking away from the age-old tradition of the circular orbit, he was not fully correct in adopting the oval as the true path of the planet. Despite partial accuracy, he took a giant step in the right direction. The correct law would be attained only when he had discovered the elliptical path. Concerning this discovery, Newton said: "Kepler knew ye orb to be not circular but oval and guest it to be elliptical."[42] A close study of our astronomer's discovery of the law reveals that although he did not arrive at the elliptical orbit by strict mathematical derivation, his process of discovery, far from guesswork, was the culmination of years of work that took him through a labyrinth of various methods and innumerable trials.

Kepler's move from the oval to the elliptical orbit can be divided into four sections: (1) the rejection of the epicyclic theory of production of the oval; (2) the acceptance of the librational theory of production of the oval; (3) the discovery of the ellipse; (4) the discovery of the relation between the librational model and the ellipse. I shall now discuss these four stages, focusing especially the first two, since they have not been sufficiently emphasized by most Kepler scholars, although Kepler himself considered the two points extremely important.

Rejection of the Epicyclic Theory of Production of the Oval

As soon as Kepler had established to his satisfaction that the orbit was an oval, he moved to find a method by which it could be produced. All through his work he showed a dual concern: to discover both the shape of the orbit and the mechanism by which it was produced. This twin concern was nothing but a natural outcome of his goal and his beliefs. Since his goal was to build a physics of the heavens, he could not be satisfied with a mere knowledge of how the orbit looked. He wanted to know what forces were involved and how they interacted to give rise to such a shape.

He believed that the techniques of chapter 39 of AN could aid him here.[43] We have seen that earlier these techniques proved fruitless because of the absurdities they engendered. However, he argued that the absurdities arose because he was trying to produce a perfect, circular motion. Now that he no longer looked for a circular orbit, he thought that such absurdities would not arise. Kepler lamentingly recounted that had he followed this line of thought "rather more wisely," he would have come to the truth. But unfortunately for him, in his haste, he fell on the epicyclic model of production of the oval, which led him to a maze of problems. Chapters 45 through 50 of AN describe how seriously this deviation hampered his work. Koyré says that for the whole of 1604 and the beginning of 1605 Kepler was lost in the mire of this deception.

The basic idea of the epicycle model was that an oval orbit could be obtained if we assumed that the planet described a uniform motion on the circumference of the epicycle, while the center of the epicycle executed a nonuniform motion on the deferent. Kepler now proposed a hypothesis according to which the inherent planetary force caused the planet to progress by equal amounts in equal periods of time and enabled it to make uniform approach to the sun in accordance with the law governing the epicycle. On the other hand, the sun's force caused the planet to advance in a nonuniform manner, according to its distance from the sun. As a result, the distance of equal arcs on the epicycle would increase towards the apsides and decrease in the mean longitudes. This could explain the increased distances at the apsides and the decreased distances in the mean positions of the orbit. "Now since this agreement [*conspiratio*] carried tremendous power, I at once concluded that the inflection of the planet towards the interior came about because the force moving the planet and determining the distances [from the sun] in accordance with the law of circularity has precedence over the sun's force. . . ."[44]

Kepler admitted that if the diameter (radius) of the epicycle had remained parallel (*aequidistans*) to the line joining the sun and the center of the circle, he could have steered clear of his erroneous view and adopted the most true opinion that every advance along the length of the zodiac was to be ascribed to the sun and only the libration along the diameter of the epicycle (radius vector) was to be attributed to the planet.[45] (This point, of course, he noted much later, perhaps after he had resolved the problems. It meant that if he had allotted a far more major role to the sun, he would have been on the right track.) At this stage he thought that this theory agreed with observation, and he persisted in his erroneous belief.

Kepler now wanted to describe the oval in the light of the ideas proposed in chapter 45 of AN, i.e., the epicyclic production of the oval. Finding that the geometrical methods were of no avail, he had recourse to the vicarious theory. (In figure 8,[46] S = the center of the sun, SI = the line of apsides, and SE = the eccentricity of the equant. Let, as in the vicarious hypothesis, SC = 11332 and CE = 7232. With C as center and CH = 100,000 let an eccentric be drawn. Let EH be joined. Here the angle IEH will be the mean anomaly, which will give the measure of the time under consideration. Angle ISH will be the true anomaly. The planet will be on the line SH since the vicarious hypothesis gives the longitude with sufficient accuracy. However, SH will not give the correct distance of the planet, because the planet will not be at H since Kepler has proved the bisection of the eccentricity and so the circle drawn about C cannot give the right position of the planet.

Let another eccentric IL be drawn with O, the midpoint of SE, as center and CH as the measure of radius. Kepler called the eccentric fictitious since he had already shown that the path of the planet was not circular. As claimed in chapter 45, the planet by its own intrinsic

Figure 8

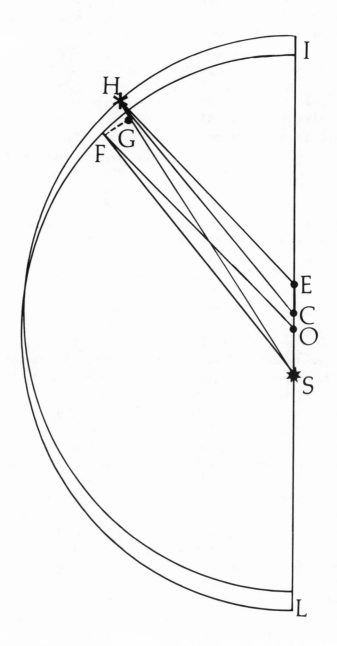

force moves uniformly around O and makes the angle IOF = IEH so that after time IOF or IEH the planet will attain its right distance SF.

With S as center and SF as radius let an arc FG be drawn, cutting SH at G. Now SG will give both the position and distance of the planet from the sun.

Kepler now concluded: "Therefore the line SG is formed by two manifestly false hypotheses, although it is true with respect to the position in the zodiac and in its length agrees with the hypothesis of chapter 45."[47] SG was determined by means of two false hypotheses: the vicarious hypothesis and the circular hypothesis. The former could fix the position of the planet, but could not give the distance; the latter could give the planet's correct distance from the sun, but not the correct position. But with the help of these two "manifestly false hypotheses," he claimed to have arrived at the correct result. He observed that in all his attempts the curve obtained was an oval and not an ellipse. "Whichever method discussed above is employed to delineate the path of the planet, it follows that the path is truly an oval, not an ellipse."[48]

In order to deduce the equation of the motion of the planet Kepler had to know not only the shape or description of the curve but also its quadrature. To determine the area of the oval he needed to ascertain the surface area of the "lune" between the oval and the circle. Since the oval was irregular, he used an auxiliary ellipse for his calculations. He found that the area of the lune was almost equal to that of the small circle formed by taking the eccentricity as the diameter. Chapter 47 of AN discusses his determination of the area of the oval. If the radius of the circle was taken as 100,000, the eccentricity would be 9264. He found that with this value of the radius, the maximum width of the lune was 858, just as, if the area of the circle was 31 415 900 000, then the area of the lune (*planum circelli*) would be 269 500 000.[49] Substracting the lune, the area of the oval came out to be 31 146 400 000.

He used the method of areas to determine the orbit, but the results were not good (see table 1).[50] These results showed that the planet was slow at the apsides and fast at the mean longitudes. For the first time, the findings prompted him to call into question the reliability of the epicyclic model of chapter 45. He remarked that either the opinion of that chapter was defective (*vitiosa*) or it was developed on the basis of faulty method.[51]

Table 1

Mean Anomaly			Vic Hyp			Oval			Difference	
°	′	″	°	′	″	°	′	″	′	″
48	45	12	41	20	33	41	14	9	−6	24
95	18	28	84	42	2	84	39	42	−2	20
138	45	12	131	7	26	131	14	5	+6	39

Seeing that the geometrical method had many problems and yielded unsatisfactory results, he decided to try the arithmetic method, despite the arduous and laborious task involved. Accordingly, he divided the orbit into small segments of one degree and calculated their distances one by one. The results this time were more encouraging (see table 2).[52] These better results gave him a feeling that he was on the right track. Hence he concluded: "Exceedingly did I congratulate myself and I was confirmed in the view of chapter 45."[53]

Table 2

Mean Anomaly	True Anomaly			Vic Hyp			Difference	
°	°	′	″	°	′	″	′	″
45	38	2	24	38	4	54	−2	30
90	79	26	49	79	27	41	−0	52
135	126	56	25	126	52	0	+4	25

Although this method gave better results and he seemed to have been more and more assured that he was getting closer to the truth, his sense of scientific honesty kept him far from being satisfied. This method suffered from a number of serious problems. For one thing, it could give a separate and independent equation only for the part of the first degree of anomaly. For all the other parts one had to know the equation immediately preceding the part under consideration. This meant that if there was any error in the previous equation, it would be retained, or even magnified. The method had another difficulty as well: "The foundation of the whole was wanting; that is, the length of the whole oval circumference was unknown."[54] The method open to him to determine the length was that of false position where one assumed the length of the oval circumference and then verified the assumption by adding together the parts. If the parts were expressed in angular measure, then they together should add up to 180 degrees to give the correct result for the semioval. Obviously he had to carry out the laborious procedure several times before he could get a satisfactory result. He reported that for each degree of mean anomaly he had to carry out this procedure three times.

The method was fatally afflicted by the malady of a vicious circle, as he himself frankly pointed out:

> It is impossible to know by this method how large a section of the oval orbit corresponds to a given time without knowing from the beginning the length of the whole oval, even if the distances of that section is known. However, it is impossible to know the length of the oval, unless the measure of the planet's inward deviation from the circumference of the circle along the sides is known. But one does not know the measure of this deviation before one can know how large a section of the oval path is traversed within a given time. Here you see the vicious circle: in our operation we have assumed in advance what we are looking for, namely, the length of the oval.[55]

Thus here, on the one hand, the length of the oval could not be determined without first ascertaining the deviation of all its points from the eccentric; on the other hand, determination of these deviations required the length of the circumference of the oval. For Kepler this was no merely philosophical or purely intellectual problem. He emphasized that the issue of a vicious circle that was involved here was most alien to the creator and ordainer of all things. This situation strongly persuaded him that there was something radically wrong with the epicyclic theory of the production of the oval espoused in chapter 45.

There was yet another problem. As Small points out,[56] the oval produced by the combined action of the solar and the intrinsic planetary forces was divided into unequal parts, the inequality being governed by its distance from the sun. This unequal division came about because of the distance law which said that the force varied inversely as the distance. But the law of variation was applicable only in the case of the solar force, the intrinsic planetary force being assumed to be invariable in every distance. This obviously raised doubt about whether the different portions of the orbit had been accurately measured.

We have a strange situation here: on the one hand, the method was beset with many defects; on the other hand, it yielded results that were much better than any of the previous ones. Kepler certainly was delighted to see the better results, but he was not willing to compromise other principles just for the sake of better results. He declared: "Even if we have come very close to the truth by means of this method, if we have been abandoned by reason, we have no right to boast. . . ."[57]

Hence Kepler decided to go back to the original principles of chapter 45. Leaving aside both the oval orbit produced by the action of the combined forces of the sun and the planet and all considerations of the quadrature, he decided to deduce the equation of motion of the planet from the following principles: that the planet moved uniformly on the circumference of the epicycle while the center of the epicycle moved nonuniformly on the deferent, the speed of motion of the center of the epicycle varying inversely as the distance of the planet from the sun. Using this method he deduced the true anomalies of the planet. The results obtained were disappointing (see table 3).[58] This obviously was worse than what he got in his previous attempts. But he refused to be discouraged. "Good friend," he addressed the reader, "if I were worried about the results, I could have dispensed with all these labors and remained content with the vicarious hypothesis. Be assured that these errors would be for us the way to truth."[59]

This failure convinced him that the source of error was not so much in the calculations or in the lack of geometrical accuracy or in the distortion or misrepresentation of the theory, but in the very theory itself. But it is surprising that despite all these disastrous consequences, he refused to part with his model of chapter 45. Even at this stage he looked for reasons to defend it. Immediately after declaring that the source of the problem was in the ideas of this chapter, he added: "This is not to say at once that the whole opinion is false, but that we have been exceedingly precipitous. . . ."[60]

Kepler now wanted to employ another method, a more elaborate and complex one, which would finally establish the falsity of the ideas

TABLE 3

Mean Anomaly °	True Anomaly °		′			Vic Hyp ′	″	Difference ′	″
45	37	56	43	38	5	0		−8	00
90	79	26	35	79	27	0		−0	25
120	110	28	8	110	18	30		+9	38
150	144	16	49	144	8	0		+8	49

of chapter 45. This method consisted in the accurate investigation of the observed distances of the planet from the sun at various angles of anomaly. As he described in chapter 51, this investigation determined 12 distances. Later on he would compare these distances with the calculated values. Kepler also investigated the distances of the planet from observations before and after its opposition when the distances of the earth from the line of syzygy were equal. This method yielded 28 distances.[61] Before comparing the observed distances with the calculated ones, he wanted to determine more accurately the elements of the orbit. As he reported in chapter 54, the values obtained were these: aphelion distance = 166465; perihelion distance = 138234; radius of the orbit = 152350; and eccentricity = 14115.[62]

He proceeded now to compare the observed values with the calculated values. He had made observations on either side of the orbit, i.e., he studied the corresponding positions of the planet in both semiovals. The anomaly of the corresponding point in the second semioval he called the complementary anomaly. The results obtained are shown in table 4.[63]

This long investigation led him to the conclusion that the path of the planet was neither a circle nor an oval. The circle was too much outside, it erred by an excess; the oval was too much inside, it erred by deficiency. According to him, the oval model erred by as many as 660 units in defect, the radius of Mars being taken as 152350.[64] He added that Fabricius also had reached a similar conclusion on the basis of observational study. This demanded that the breadth of the lunula cut off by the orbit from the circle must be half of what was given on the basis of the oval theory.

Reflection on his findings led Kepler to an "accidental," but very important, discovery. He noticed that the breadth of the lunula was

TABLE 4

Mean Anomaly			Observed Distance	Complry. Distance	Calculated Distance	Difference		
°	′	″	A	B	C	C − A	C − B	Mean
11	37	0	166194	166230	166179	−15	−51	−33
43	23	31	163100	163028	162895	−205	−133	−169
70	55	0	158100	158217	157530	−570	−687	−628
87	9	24	154387	154400	153697	−690	−703	−696
113	0	0	147760	147591	146928	−832	−663	−747
162	0	0	138954	139000	138991	+37	−9	+14

just half of what was predicted by the oval theory. He also noted that if the radius of Mars was taken as 100,000, the breadth of the lunula at an anomaly 90 degrees must be 429. This whole result disturbed him greatly since his theories so far could not explain why this was so. Without getting an explanation for this halving of the width of the lunula he felt that his "triumph over Mars would be futile."[65] He was convinced that there must be a reason for everything. At this painful juncture, he says, by chance (fortuito) he noticed the secant of the angle 5 degrees and 18 minutes which was the measure of the largest optical equation,[66] i.e., corresponding to an eccentric anomaly of 90 degrees.

> When I noticed that it [the secant of the angle 5 degrees and 18 minutes] was equal to 100,429, I felt as though I had awakened from sleep to perceive a new light. I began to reason as follows: in the mean longitude, the optical part of the equation has its maximum value. In the mean longitudes, the lune or the shortening in the distances is the greatest. Now this decrease is just as large as the excess of the secant of the largest optical equation 100,429 over the radius 100,000.[67]

At the mean longitudes the breadth of the lunula was only 429, rather than 858, the value required by the oval theory. From this reasoning he concluded that if in the mean longitudes he took the length of the radius rather than that of the secant, he would get the result conformable to observation. This conclusion was only for the eccentric anomaly of 90 degrees. But instantly he generalized this finding: "I shall draw, therefore, the general conclusion: if in the scheme outlined in chapter 40 we take HR instead of HA, VR instead of VA, EB instead of EA, and so on, we shall have the same result for all other places on the eccentric as that which has been obtained here for the mean longitudes" (see figure 9).[68] This meant that if one took the diametral distance, i.e., the projection of the planet–sun distance onto the diameter through the position of the planet, rather than the circumferential distance, i.e., the planet–sun distance, then one could get a result in agreement with observation. This generalization was a big jump. At this stage he did not give any empirical justification for it. Kepler was fully aware that this generalization needed justification. He confessed: "However, what I concluded as generally applicable to all cases on the basis of one instance of the anomaly, did not as yet follow from such a one. Hence it was necessary to establish the conclusion by many more observations."[69] Therefore he wanted to investigate the matter for other positions of the planet as well. Using the observational data of chapters 51 and 53 he got the following results (see table 5).[70] Since the differences were not considerable and showed no regular pattern, Kepler believed that they were due to human and instrumental errors in observation. This result convinced him that he was justified in generalizing the result of the "accidental" discovery.

TABLE 5

Mean Anomaly			Distance Ascending semicircle	Distance Descending semicircle	Distance Calculated	Difference	
°	′	″	A	B	C	C − A	C − B
11	37	0	166194	166348	166228	+34	−120
43	23	31	163022	163100	163160	+138	+60
70	55	00	158101	158217	158074	−27	−143
87	9	24	154400	154278	154338	−62	+60
113	0	0	147760	147931	147918	+158	−13
162	0	0	139000	138984	139093	+93	+109

Figure 9

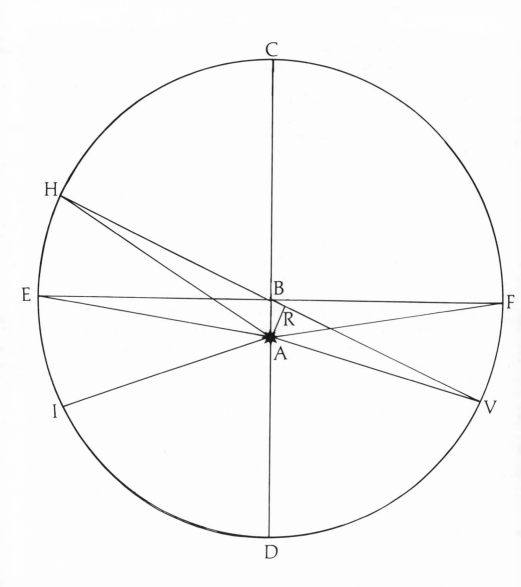

Acceptance of the Librational Model
of Production of the Oval

The investigation above convinced Kepler that the epicyclic theory of the production of the oval was false. The planet was not moving in an epicycle the center of which moved on the deferent, but, rather, the planet executed a librational motion. Even at this stage he had not given up the oval. The change we are talking about refers to the mechanism of production of the oval. He adhered to the oval strongly, as was evident from his letter to Maestlin, written on March 5, 1605: "The distances from the sun are not in a perfect circle, but in an oval. Only after infinite effort did I find the representation of this oval."[71] In this letter he explained how he came to reject the epicyclic theory and embraced the librational theory. The same idea he expressed in AN also. Thus as soon as he had made the "accidental" discovery about the need for substituting the diametral distances for the circumferential ones, he commented: "Let the reader go over chapter 39 once again. Then he will discover that what I already argued there on the basis of natural reasons, the same I have confirmed here from observations, namely that it was conformable to reason that the planet should perform a certain libration as if on the diameter of an epicycle, perpetually directed towards the sun."[72]

In figure 10, S is the position of the sun. The different positions that the planet would occupy on the circumference of the epicycle, if it were to execute an epicyclic motion are g, d, e, z, a, and n. But Kepler argued[73] that the rule of substituting diametral distances for circumferential ones would mean that in the diagram we take Sg, Sk, Sm, and Sz for Sg, Sd, Se, and Sz. Hence the planet would not be moving along the circumference of the epicycle but, rather, would be librating on the diameter of the epicycle.

The discovery of the librational theory was no easy task for him. As he put it, "it was a laborious task to reduce two false hypotheses into a true one. . . ."[74] This statement is rather surprising since here he seemed to subscribe to the view that a true conclusion could come from false hypotheses, a view he vehemently opposed earlier. (This point we shall discuss later.) He continued: "I made a thousand attempts . . . the solution was possible only by investigating the [real] causes in nature."[75] This investigation of the real causes

Figure 10

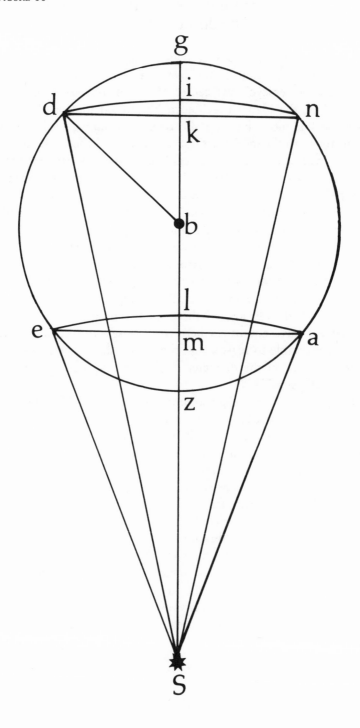

in nature was based on two major points: the supreme importance of the sun and the extreme emphasis on the need for explanation in terms of natural, real forces. AN's long chapter 57 detailed this investigation.

One main characteristic of the epicyclic theory of the production of the oval was that the sun had only a limited role. In fact, the sun had no role in the approach and recession of the planet during its motion along the orbit. This approach and recession was attributed to the intrinsic planetary force. All that the sun did was to cause the planet to advance in a nonuniform manner along its orbit. The sun played only a partial role in the actual motion of the planet. Later, when he was writing his final version of AN, Kepler regretfully added: "Thus I got confirmed in my error, which in chapter 39 I had happily begun to refute, that it was the function of the planet's own force to guide it on an epicyclic path."[76]

Kepler was convinced that the sun should play a far greater role in the motion of planets. Hence he wanted to explore what role the sun played in the librational motion of the planet. He expressed this intention right at the beginning of chapter 57 of AN: "Since, as we know on the basis of the best reasoning presupposed in chapter 39, a planet cannot move from one place to another by virtue of the struggle of the internal [*insitae*] forces alone, unless it is aided and empowered [*informare*] by an external force, it is for us to consider whether in some way we can attribute this libration in part to the solar power."[77] I suggest that one of the main reasons why he accepted the libration theory as the true one was that it assigned to the sun a most prominent role in the production of the correct orbit of the planet.

To bring out better the role played by the sun in the production of the true planetary orbit, Kepler drew a parallel among the librational, the epicyclic, and the boat motion.[78] Earlier Kepler had argued that the substitution of diametral distances for circumferential ones would mean that the planet did not move along an epicycle but librated along its diameter (see figure 11). He now considered the motion of a boat in a circular river CDEFGH (see figure 11c). The sailor could ride in the river by turning his oar in the appropriate way. The motion was such that by the time the planet made a whole revolution the oar made only half a revolution. The circular river was formed by the emanation from the rotating sun. The force inherent in the planetary body produced the twisting and turning of the oar. According to the analogy, at C the oar was perpendicular to the line of emanation from the sun and at F parallel, the angles of inclination being intermediate in the region between C and F. Kepler believed that the greater the inclination of the oar to the line of flow of the river, the greater the surface it exposed to the flow and the more it was affected by the force of the current. Hence in the region DE the effect of the flow on the boat would be significant. In this region the river flowing over the oar would press the boat down towards S, the sun. Exactly the opposite would be the case in the region GH, where the boat would be pushed away from the sun by the river flowing under the oar. On the other hand, at C and F the flow would have scarcely any effect on the boat. Kepler was of the opinion that everything else being equal, the river would be slower at C than at F and hence the impulse would be correspondingly weaker at C than at F. He claimed that the boat analogy could explain the observed characteristics of the motion of the planet.

Kepler himself admitted that this analogy suffered from serious defects. First of all, the time period of rotation of the oar was twice that of the revolution of the boat in the circular river. This meant that the planet should present continually changing faces towards the sun and the earth. Since the moon's motion was not different from that of the planet, if this analogy was correct, we should expect its face to change constantly. However, we know that the moon always presents the same view. Another disanalogy stemmed from the fact that the river was something material, whereas the sun's emanation was nonmaterial. Moreover, the planets he knew were round bodies which neither had nor needed oars.

Despite these problems, he unhesitatingly moved to magnetize the analogy. "The body of the sun is circularly magnetic and it rotates in its place."[79] The river was "the immaterial emanation of a magnetic power in the sun."[80] Thus a circular "magnetic river" was produced. At this stage he argued that "the bodies of the planets [are not in themselves endowed with motivity, but] are inclined to remain at rest in whatever part of the universe they were placed."[81] Here he had made a significant change. For in AN[82] he had believed that the individual planets did have a force, and that it was responsible for the planet's progress by equal amounts in equal periods of time and caused the planet to perform uniform approach to the sun in accordance with the law governing the epicycle. Now he seemed to

FIGURE 11

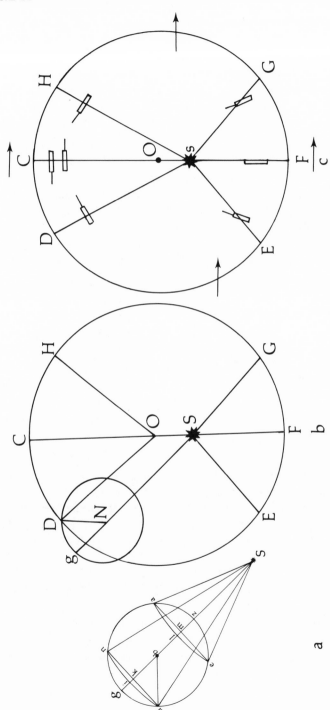

attribute to the planet only an inertial quality: it just resisted motion because of its tendency to preserve its state of rest. In this important move also we can see the influence of the two key points of Kepler's thought: the unique importance of the sun and the emphasis on physical explanation in the investigation of planetary phenomena. According to this shift in thought, the sun had a far more vital importance in the motion of the planet than had been admitted earlier. The sun had a role to play in every aspect of the planetary motion, including the planet's approach and recession from the central body. This move also made it possible to explain the phenomenon without resorting to animal forces or intelligences.

Although a planet did not have any inherent animal forces, it was characterized by a quasi-magnetic property. According to Kepler, "every planetary body must be regarded as magnetic, or quasi-magnetic; in fact, I suggest a similarity, and do not assert an identity."[83] By means of this magnetic property he explained the oval orbit of the planet. Thus the planet had two quasi-magnetic poles, one "friendly to the sun" (*soli amica*) and the other "hostile" (*discors*). The former tended to pursue (be attracted to) the sun, while the latter tended to retreat (be repelled) from it. The axis of the planet was kept constantly directed approximately to the same part of the universe by an animal force. Later he would replace this animal force with a magnetic power.

From figure 12 it is clear that Kepler considered the two paradigms, i.e., the planet as a boat and as a quasi-magnet, very analogous. At C and F since both poles are equidistant from the sun there is neither "attraction" nor "repulsion." In the region DE the "friendly side" faces the sun and so the planet will be "attracted" to the sun and continue to approach the sun until F. From F on, the "hostile" side will face the sun and the planet will be "repelled" from the sun and the distance will be increased until the planet reaches the aphelion. He declared that this theory could account for the observed planetary phenomena.

> It is attested by observation that the planet performs libration, and particularly that during libration it moves slowly in the vicinity of the apsides of the epicycle, and more quickly in the mean positions; whereas in its *raptus* round the sun, it moves slowest at aphelion, and quickest at perihelion. Furthermore, the superior

Figure 12

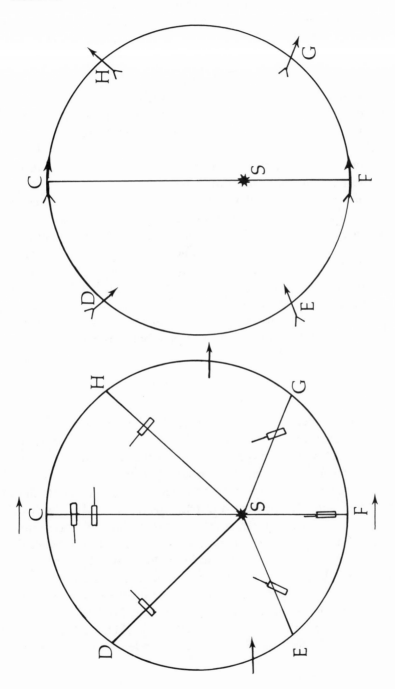

semi-diameter of the libration [upper in the figure] is traversed in a longer time than the equal inferior semi-diameter [lower in the figure]; for the magnetic force of the planet itself acts also less strongly when the planet is remote from the sun.[84]

In his letter of 1605 to Maestlin, Kepler talked of an animal force responsible for orientating the axis in the right direction. But in AN chapter 57, he wanted to explain it in terms of physical, corporeal factors. As he put it, "in simple words, since this [remaining always in the same] direction is more akin to a state of rest, rather than of motion, its cause must be rightfully sought in matter and in the composition of the body, rather than in some mind."[85] As far as the specific and particular points were concerned, there remained difficulties: for instance, he said that the earth's axis was unsuitable for this kind of libration because its axis diverged from the one of the sun's apsides.[86] However, he concluded: "In this way the librational approach is brought about by a magnetic force which is indeed innate and solitary, without any operation of a mind, but its description depends on the external solar body. The force, in fact, is defined as sun-seeking or as sun-fleeing."[87] With this long investigation Kepler was fully convinced of the truth of the librational theory.

The Discovery of the Elliptical Orbit

It seemed that Kepler had achieved the goal he had set for himself. He had obtained an accurate equation for planetary motion and discovered the method that would yield the correct distances of the planet. Hence it seemed that he could really congratulate himself for his achievement. However, he made a mistake in computing the distances and the apparent and true positions of Mars by his new method of areas. Consequently, the curve obtained was distorted, and he found that the positions in the upper part of the orbit deviated by 5 minutes and 30 seconds and in the lower part by 4 minutes. He attributed the errors to libration and even abandoned this theory for some time, which prompted him to take up the ellipse as a serious option. Kepler summarized these ideas and his final identification of the true orbit of Mars as an ellipse in AN, chapter 58: "My argument has been as follows: . . . the circle of chapter 43 errs by excess and the ellipse [i.e., the auxiliary ellipse used to facilitate computation] of chapter 45 errs by deficiency. The excess and the deficiency are equal. But only another ellipse can be placed mid-way between the circle and ellipse. Therefore the path of the planet is an ellipse."[88] The same idea he had conveyed to Fabricius earlier in a letter dated October 11, 1605. In his own words: "Therefore, I have the answer, my dear Fabricius; the path of the planet is a perfect ellipse . . ., or deviates from it only by a very imperceptible amount."[89] Later in chapter 59 he would establish the elliptical orbit by a geometrical method.

The Discovery of the Relation between the Planetary Libration and the Elliptical Orbit

Kepler was convinced that the planet performed libration. At the same time he believed that the orbit of the planet was an ellipse. Furthermore, the libration theory gave the correct distances, whereas the ellipse with the area law yielded the correct equation. However, he had no idea about the interconnections between the two. He said that he could not understand why a librating planet should choose to move along an elliptical path. Again, he could give a physical explanation for libration, but not for the ellipse. This situation was most distressing to him. As he himself put it, it drove him to the brink of insanity. Fortunately, later he did discover the relationship. He could prove that the two were equivalent. In fact, as Aiton points out, in a flash of insight, Kepler found that a perfect elliptical orbit could be produced by the libration theory, if the planet was placed on the perpendicular from the position on the eccentric to the line of apsides.[90] He established exactly this relationship in chapter 59, which carries the title: "Proof that the orbit of Mars, performing libration on the diameter of the epicycle, becomes a perfect ellipse. . . ."[91]

Factors Responsible for the Move from the Oval to the Ellipse

With the discovery of the elliptical orbit, Kepler reached the high-point of his new astronomy. I have gone into considerable detail in the discussion of this final step in Kepler's work on Mars because some points have not been stressed sufficiently by Kepler scholars, not even by Small and Koyré. Small, for instance, almost bypasses Kepler's magnetic theory of the motion of the planet.[92] Koyré passes over several of Kepler's important empirical investigations. Our study shows how far from the truth Newton was when he wrote that Kepler arrived at the first law by guesswork. Nor did our astronomer arrive at the law by Keostlerian sleepwalking. Nor, again, did he get it by meticulous curve-fitting. Indeed, his discovery traversed a bit of all these different paths. But he did not obtain, nor could he have obtained, the law by means of any single one of these means. Koyré traces the secret of Kepler's success to his emphasis on astronomy as a science of reality and his insistence on physical explanation in terms of real forces.[93] As through all this book, I shall argue that not only empirical evidence, but also philosophical convictions and religious beliefs must be taken into account in attempting to explain how Kepler arrived at his all-important conclusion.

SCIENTIFIC OR EMPIRICAL IDEAS

The role of empirical inquiry in the discovery is evident from the detailed study above. We find Kepler time and again using the accurate data of Tycho in so many different ways. However, it is evident that if our astromoner had relied only on observation, he would not have been successful. First of all, Kepler had a dual concern: to find out the nature or shape of the orbit, and to explain in terms of real forces how this shape was brought about. Both were extremely important for him, but what marked his brand of astronomy from the work of most of his predecessors and contemporaries was the second concern. In a significant way, explanation in terms of real forces was what made his astronomy truly Astronomia Nova. This aspect could not have been achieved by relying on empirical factors alone. His own words about the discovery disclosed this fact. Remarkably, he did not say that he got the ellipse by plotting his numerous coordinates of the planet's position; he could claim a large number of such positions very well worked out.[94] Yet his key argument in the letter to Fabricius and in chapter 58 of AN was based on the inadequacy of both the circle and the oval to represent satisfactorily the path of the planet. One may say that this was an argument based on empirical data and hence Kepler arrived at this conclusion on the basis of empirical data. This argument has several problems. All that Kepler could say on the basis of observations was that the path should be in between the circle and the oval. One could have a different oval there. Tycho's observations did not tell Kepler that he should rule out this possibility and adopt the ellipse. They could not conclusively show that the orbit, the shape, was an ellipse. Furthermore, even if they did, they could not have explained how such an elliptical path was produced, nor could they bring out the crucial action of the sun on the motion of the planet.

PHILOSOPHICAL IDEAS

I have already pointed out Koyré's remark that philosophical considerations had an influence on Kepler's discovery. This influence was pronounced in Kepler's second concern of determining the mechanism by which the true path of the planet was produced. In this part of Kepler's work four philosophical principles in particular were quite active: geometrizability, causality, rationality, and unity.

Although all that observations could establish was that the path of Mars should be in between the circle and the oval, if the universe is geometrical, in the sense that God created it in accordance with geometrical patterns and laws, then there is good reason to believe that the path of the planet should have a regular geometrical shape. Thus when one adds this geometrical principle to the conclusion arrived at on the basis of observation, the argument in favor of the elliptical path becomes much stronger. Since Kepler was a firm believer of this principle, it must have been fully operative in his discovery of the elliptical path.

The principle of causality also was active in his studies at this stage. He was convinced that the universe is causally connected. In his view, the causes responsible for natural phenomena were real causes, physical causes. He believed that natural phenomena should be explained in terms of physical, tangible forces and corporeal bodies. Search for real forces and explanation in terms of them was one of the main goals of his science. This belief in the principle of causality formed the basis for his insistence on the determination of the mechanism by which the orbit of the planet was produced.

Belief in the rationality of nature was another element playing a significant role in his work. He believed that nature did nothing without a good reason and that reason was discoverable. As we have seen, this principle was so important for him that not even the fact that he got better agreement with observed data could persuade him to compromise this principle.[95] According to him, no theory that failed to honor this principle was worth pursuing.

Perhaps the role of the principle of rationality of nature was most conspicuous in the case of the "accidental" discovery, for one of the principal reasons why Kepler could make this discovery was his belief in this principle. The discrepancy between the theoretical and observed values of the width of the lunula was very disturbing and unsettling for him. As he described the situation: "See how miserably I tremble before the discovered truth. . . . My old ellipse [i.e., the oval where he had used an auxiliary ellipse] with a shortening of 0.00858 had a natural cause. . . . On the other hand, if the ellipse had a shortening of 0.00429, then it lacked any natural cause."[96] Of course, he admitted that the old oval was not fully satisfactory. Yet it had a natural cause, whereas the new one lacked any kind of natural cause. This situation created by the new width of the lunula was serious and threatening. Yet he did not give up because he believed that since nature did nothing *temere*, at random, there would be a discoverable reason for it. We know that his belief did bear fruit and he could come up with the law of substituting diametral distances in place of circumferential ones.

If he had not subscribed to the principle of rationality of nature, he could have taken one of the two paths: either give up the whole enterprise because it seemed to be leading nowhere, or revert back to the circular path. Neither of these choices was unjustified. In fact, Fabricius chose the second option. He had accepted the possibility of the oval, but when he saw that the width of the lunula had to be much smaller than that predicted by the oval theory, he concluded that the oval was wrong and became all the more confirmed in his adherence to the circular orbit. Thus Kepler's commitment to the principle of rationality of nature helped keep him from giving up his search or from reverting to the circle theory. Furthermore, this principle encouraged him to forge ahead to make significant progress in his work.[97]

The principle of unity of nature greatly assisted him in his attempt to give a satisfactory (at least, at that time) physical explanation for planetary motion. He at first developed the theory of libration of the planet. Then he compared it to the motion of a boat across a flowing river. This was an attempt to give a physical explanation for planetary motion in terms of known or knowable, real forces and corporeal bodies. However, he found that this analogy had serious problems, e.g., water is material, whereas the emanation from the sun was supposed to be nonmaterial. The boat required oars and rudder, whereas the planets neither had nor required such things. At this juncture Kepler made an ingenious move to relate the boat analogy to a magnet analogy. He hypothesized that the planet was a giant magnet and that the sun behaved like a spherical magnet with one polarity on its entire surface. He argued that the two analogies bore very close resemblance. Both could explain libration quite well, despite certain problems. This relating of the boat motion with the magnet motion had a very salutary result: the boat analogy involved familiar bodies and forces, but it had serious drawbacks when used as an explanation for planetary motion. The magnet analogy was free from these problems, but it had some other problems, e.g., if they were true magnets, then the sun and planets should attract each other and eventually collapse onto each other. Again, Kepler was not sure of the magnetic properties of the planets and of the sun (note that he referred to them as quasi-magnetic). However, the way he handled the two analogies had the beneficial effect of drawing from the strengths of both and reinforcing the positive analogies, at the same time as eclipsing their defects.

Kepler had stipulated that an acceptable physical explanation must be in terms of known or knowable, real forces, as opposed to fictitious ones, and in terms of corporeal bodies. Planetary magnetism was rather mysterious at that time. (It was only in 1600 that Gilbert published his *De Magnete* in which he argued that the earth was a magnet, but the magnetic nature of other heavenly bodies was still not adequately explored.) But the relationship with the boat analogy helped take the mystery out of the case. The boat analogy had problems with the exact nature of the forces,[98] but the magnetic analogy helped get around this problem since the magnetic forces were considered to be immaterial. Thus when both analogies were put together, by drawing a close parallel between them and by arguing that the one could explain the phenomena just as effectively as the other, the difficulties seemed to fade out and Kepler could claim to have offered an acceptable physical explanation for planetary motion. In this way he used these two analogies in a piecemeal fashion.

How could Kepler connect the mechanical forces involved in the boat motion with the magnetic forces? The two were different and the close connection between them had not been established. The First Law of Thermodynamics, the Law of Conservation of Energy, was still more than two centuries away. What persuaded Kepler to believe that boat motion and magnetic forces were intimately interrelated and hence both could be used to explain the same phenomenon? I think that Kepler could unhesitatingly and, as it were, automatically make this interrelation because he believed in the principle of the unity of nature. Everything in nature is interrelated, the different forces in nature are interconnected, and an inner unity exists among them. Hence the use of one force to explain a phenomenon does not exclude the simultaneous use of another to explain the same phenomenon. What the one lacks, the other can supply without any incongruity or conflict. In this way the principle of unity of nature aided him in his attempt to develop his physical explanation for planetary motion.

The influence of the principle of unity was evident also in his insistence on finding out the relationship between the planetary libration and the elliptical orbit. He knew that both libration and ellipse explained different aspects of planetary motion: the libration theory gave the correct distances, whereas the ellipse theory helped to get the correct equation. His firm belief in the unity of nature convinced him that these two must be related. He assiduously sought for this relationship and his efforts were finally successful. We know that the discovery of this relationship was an essential step in Kepler's discovery of the first law.

The influence of philosophical principles on Kepler's work at this stage was by no means insignificant. Without this influence, he would probably have remained with the epicyclic theory of production of the oval. After all, he himself said that the epicyclic theory could account for the observed characteristics of the motion of the planet.[99] But this model could not satisfy his philosophical and, as we shall see shortly, his religious principles. Hence he rejected it and accepted the librational model which eventually led to the first law.

RELIGIOUS IDEAS

Although Koyré and others have recognized the influence of philosophical ideas on Kepler at this stage of his scientific work, Kepler scholars have been silent about any influence of his religious beliefs on the discovery of the ellipse. They seem to imply that there was no such influence. However, a careful reading of his chapters of AN and his letters during this period reveals that even at this time the influence of his religious beliefs on his scientific work was quite significant. As we have discussed in the introduction, explicit reference to religious and related principles had become less frequent at this stage of his work. But we have good evidence to believe that this did not signal a "Comtean death of religion" in his work. Clearly God and religion did not disappear from his worldview, but, rather had become too well-assumed or taken for granted to need explicit mention.

That religious principles were active in Kepler's thoughts at this stage of his work is quite evident from his correspondence. As we have seen, Kepler announced the discovery of the ellipse for the first time to Fabricius on October 11, 1605. Only a few months before this letter he had written to Maestlin describing how he arrived at a major breakthrough in his work on Mars: "Oh, the immense and most perplexing labor involved. But with God's grace I conquered it."[100]

Again, that the extraordinary, divinelike importance he attached to the sun was as powerful as ever in his mind at this period is affirmed by another letter written even later and so still closer to the announcement of the ellipse.[101] This letter to Herwart highlighted the Trinity–sphere analogy and the supreme importance of the sun.

Not only during the period of the discovery, but also soon afterwards and in later years, religious principles were fully alive and active in his thought and work. We find evidence for it in *Tertius Interveniens*, section 126, published in 1610, soon after the publication of AN. Similar ideas can be found in his *Dissertatio cum Nuncio Sidereo*, also published in 1610. We have seen that MC and HM emphatically proclaimed the centrality of God and religion in all his thinking and work. He made this point undeniably certain in the dedicatory epistle for the second edition on MC, published in 1621: "Almost every chapter on astronomy which I have published since that time [i.e. 1596, the year of publication of the first edition] could be referred to one or another of the important chapters set out in this little book, and would contain either an illustration or a completion of it."[102] Undoubtedly, religious principles were active in him before, during, and after his discovery of the law of the ellipse.

More specific considerations also argue for the role of religious principles in this context. We have seen in this chapter that Kepler insisted on identifying the mechanism of production of the orbit. He did so because he believed that the aim of science was not only to "save the phenomena" but also to explain them. A complete explanation should include the "how" aspect of the planetary orbit. Kepler's insistence had a strong religious basis as well. For him, scientists were "priests of God" who expounded the "deed of God." The goal of science was to uncover the plan of God expressed in the laws and workings of nature. Such an uncovering of the great plan of God would remain incomplete if the scientist failed to go beyond the appearances of the planetary path to investigate the real forces and other causes responsible for the observed orbit.

We have seen that both Kepler's rejection of the epicyclic model of production of the path of the planet and his acceptance of the librational model were important for the discovery of the ellipse. Kepler's religious beliefs made contributions to both of these decisions.

We find him calling into question the reliability of the epicyclic model, precisely when it seemed to yield better agreement with observation. He noticed that use of his method led to the fallacy of vicious circle. He considered this case totally unacceptable because he believed that such a "vice was most alien to the primeval ordainer of planetary courses. . . ."[103] According to him, such a state of affairs went against the mind of God, the prime mover and ordainer of the heavenly bodies. Hence a theory giving rise to such a position could not be correct. The verdict of observational data alone would have misled him and prevented him from going ahead to make his discovery, but his conviction that the omniscient and most rational God could not allow such a situation helped him break out of the epicyclic trap. It is remarkable that this was one of the rare instances where Kepler was willing to discount the verdict of observation because philosophical and religious beliefs opposed it. Thus even at this developed stage of his empirical inquiry, when he seemed to have reached the pinnacle of his career as a scientist, he was unwilling to give up his philosophical and religious principles.

We have seen also that Kepler successfully used the magnetic hypothesis in his investigations and thus got rid of the spirits, minds, and souls from physical explanation. But, as I have pointed out, there were problems associated with this hypothesis. What was it that encouraged him to accept this magnetic power in place of the minds, despite the problems with the former? In the same chapter he wrote that the use of the mind in this case involved a certain geometrical uncertitude. Such an uncertitude he believed would be "repudiated by God . . . who up to now is understood to proceed always by the path of demonstration."[104] The explanation in terms of mind was not acceptable because it led to serious geometrical uncertainties in our knowledge of nature. Since God the perfect geometer was the creator of the universe and God created it on definite and certain principles, an explanation which yielded uncertain answers could not be worthy of God, and therefore could not be true. Hence explanation in terms of minds was unacceptable. Clearly, Kepler's religious beliefs had a role to play in persuading him to banish the spirits and souls from scientific explanation.[105]

An interesting observation about the discovery of the first law is that although Kepler knew that his theory of planetary libration had serious problems, he accepted it wholeheartedly. The theory had a rather sinister origin. As he himself admitted, it was formed by the reduction of two false hypotheses into a true one.[106] In MC he had vehemently rejected such a possibility: "On this point I have never been able to agree with those who rely on the model of accidental proof, which infers a true conclusion from false premises by the logic of the syllogism. The conclusion from false premises is accidental. . . ."[107] Since for Kepler nothing happened in the universe by accident, accidental was synonymous with unreal. The same point he made in the *Apologia*:

> Certainly one should not suppose that it happens because what is true customarily follows from what is false. For, as I said before, in the long and tortuous course of demonstration through diverse syllogisms of the kind which customarily occur in astronomy, it can scarcely ever happen, and indeed no example occurs to me, that starting out from a posited unsound hypothesis there should follow a result altogether sound and fitting to the motion of the heavens or a thing of the kind one wants demonstrated.[108]

Kepler's original position, therefore, had been that a true conclusion could not come from false hypotheses or premises, except in an accidental, artificial, and hence unreal way. The true nature of reality or a true phenomenon could not be established on the basis of false hypotheses. Yet he seemed to be saying exactly this in the case of the origin of the libration theory.

Again, his use of the magnetic hypothesis to explain the libration also involved a problem. Kepler himself admitted that he was talking in general terms. When one went into specifics, there were difficulties. As we discussed already, he thought that the case of the earth did not fall in line with his position.

I think that one of the main reasons why he accepted whole-heartedly the libration theory and the magnetic explanation of it, despite all these openly acknowledged difficulties, was that this move assigned a far greater role to the sun, a role that seemed to have been seriously diminished in the epicyclic theory. He was deeply dissatisfied with the latter theory since it gave only a partial role to the sun. As he himself announced in the beginning of chapter 57 of AN, the whole chapter was aimed at seeing whether a far more important role could be attributed to the sun. To his great joy and satisfaction, he found that this indeed was the case; this libration was really dependent on the sun. In fact, the forces involved were defined as "sun-seeking" and "sun-fleeing," thereby accentuating the supremely important role of the sun. Hence the fact that the libra-tional theory could affirm such an unparalleled role for the sun could be considered as one of the principal reasons for Kepler's acceptance of the librational theory as true, despite troubling problems.

We have seen several times the religious basis and significance of this role. Kepler always wanted to build a God-centered science and the placing of the sun as the center of activity, as the unique and unparalleled cause of planetary motion giving rise to the true elliptical orbit, could and did assure such a central role for God. Thus his religious beliefs once again contributed significantly in reaffirming the special role of the sun.

All three factors, empirical, philosophical, and religious, were nec-essary for Kepler's discovery of the first law. Philosophical and reli-gious ideas were not mere members of a "rescue squad" but "respectable partners." The empirical data established that the path of the planet must be mid-way between the circle and the oval. The philosophical realm, with its insistence on geometrical laws, could specify that the path had to be an ellipse and could also help discover the mechanism by which the orbit was produced. Religious beliefs could give a firm grounding for the philosophical principles used and, above all, could ensure that the sun was given its due prominent role in producing the path. All three areas of ideas were needed to lead Kepler to final victory in his venture to build the New Astronomy.

CONCLUSION

AN INTEGRATED VIEW OF
KEPLER'S THOUGHT

My study has presented a vivid and detailed picture of Kepler's long journey toward the discovery of the laws of planetary motion, and I identify the various factors involved in this discovery and analyze how each factor contributed to final success. One of the most difficult tasks in this study is to locate and distinguish the religious, philosophical, and empirical ideas that shaped his thought and works. To my knowledge, such a study has been regrettably absent in Kepler scholarship, at least in the English language. Similarly, a detailed in-depth study of the various steps explaining how each of the different factors concurred to make the discovery possible also has not been attempted. Koyré and, before him, Small have discussed in great detail the various steps. But not only do they fail to identify all relevant major factors active in this process of discovery, but also the authors fail to discuss the interaction of those factors. My study aims at remedying these deficiencies, by presenting a considerably more adequate picture of Kepler's thought and work, as far as the discovery of the two laws is concerned.

246

My discussion reveals the complexity of scientific discovery in general, and Kepler's discovery in particular. It supports my thesis that the discovery of the laws required the active participation of all three factors: empirical (scientific), philosophical, and religious. In some stages one factor was more prominent than in some others,[1] but all three were present and made unique contributions throughout the various stages. Therefore, any attempt to depict Kepler's work as a simple, straightforward inference on the basis of Tycho's data, though enviably abundant and admirably accurate, will be futile. Strong and other positivist-minded philosophers will try in vain to see in Kepler a positivist who had to banish nonempirical factors in order to make progress in his science. Describing Kepler's discovery as the glorious outcome of a process of curve-fitting does nothing but betray ignorance of Kepler's own account. On the other hand, characterizing the Keplerian achievement as the paradigm of "sleep-walking" also fails to tell the true story. Indeed, there were unpredictable elements in his work, but, as we have seen, they were not totally surprising. There were uncategorizable elements in his thoughts and ideas, yet they were governed by certain rules of rationality. This study, thus, attempts to avoid the deficiencies that have plagued previous research on Kepler's thought. Earlier studies remained either inaccurate or incomplete because they failed to take into account all the important factors involved in Kepler's work.

Throwing light on the role of what we today consider "extrascientific" factors in scientific discovery, this investigation shows that any attempt to decouple Kepler's work from them is bound to be futile and incorrect. At the same time, my work reveals the untenability of the view which wants to overemphasize the importance of these factors, as Burtt did when he argued that Kepler's "sun worship" was the sufficient reason for his acceptance of heliocentrism. In this regard, too, my position avoids a one-sided or extreme view, by giving due credit to all the elements involved. By emphasizing the constructive intermingling of the empirical, the philosophical, and the religious, and by specifying how they collaborated with and complemented each other, I hope I have provided a picture of the characteristic Keplerian method of science which has been largely missing from previous studies. Indeed, what I have tried to offer is a truly integrated view of Kepler's thought, a view in which the different factors, which some scholars have distinguished in order to show their irrelevance, are shown to play crucial roles.

One of the significant points that emerge from this book is that Kepler's philosophical and religious ideas were not marginal or subsidiary but played a vital role, indeed a unique role that they alone could have filled. In Kepler's work philosophy and religion were not just called in to meet an emergency; rather, they collaborated in a common mission. Although this idea is quite clear in the various steps of Kepler's discovery of the laws, it has not been brought out adequately by any Kepler scholars before.

One may raise the objection that this study also shows that in Kepler's work the empirical realm predominated, and the philosophical and especially the religious areas had only minor or ancillary roles (making the latter two only a "rescue squad," after all). It is possible that science and observation supplied more ideas than philosophy and religion, but this does not rule out the need for ideas from those two latter domains. What I argue for is that in every step the religious, just like the philosophical, had to play a specific role to make the successful conclusion possible. They were needed for guiding and initiating crucial ideas and their contribution, not reserved to rare, emergency situations, was needed at every step: we are talking about normal situations and Kepler's most ordinary, typical processes of discovery. The objector confuses the questions Whether each one is needed? with How much is each needed?

Nor were the philosophical and religious a mere scaffolding, temporary planks set up only to assist the empirical in a specific task without having any lasting and enduring place in the whole project. I have shown that, though in varying degrees, these two were always present, even at the height of his "Tychonism." They had a substantive and structural, rather than an accidental and ornamental, place in Kepler's discovery.

My work attempts also to clarify and complete points that have hitherto been left vague and incomplete. For instance, although eminent scholars like Wilson had talked about the role of certain "hunches" in our astronomer's work, their nature and source were left undeveloped. Our study has unpacked these Keplerian "hunches" by tracing their sources to his religious and philosophical commitments. It has shown that the presence of these "hunches" did not render the process of discovery irrational, because they had a rational foundation; they did not arise in a (rational) vacuum. Thus Kepler had a "hunch" about the constancy of the inclination of the planetary orbital plane to the ecliptic. I have shown that this "hunch" had its source in Kepler's philosophical principles of economy and simplicity, and in the extreme importance he attached to the sun. The special place of the sun in his system had its roots in his religious beliefs.

This study also brings out the originality of Kepler's methodology and its striking similarity with certain contemporary views. As we have noted, Kepler's so-called a priori method had all the essential elements of the H–D method, with the exception that he believed that the hypothesis had a real rational basis. In this emphasis on the rationality of hypothesis formation Kepler was surprisingly close to certain dominant contemporary views of the philosophy of science. In contrast to the earlier view, several present-day philosophies of science hold that the initial hypothesis in the H–D method is not a product of pure chance or the result of irrational psychological activity. Kepler's philosophical principle that nothing in nature happens by chance holds good in the formation of the initial hypothesis as well.

Thus our investigation also provides insights valuable for the logic of discovery and reveals that Kepler's discovery was neither an irrational, purely psychological process nor a process governed by strict rules of any system of logic. His case renders both the Popperian and the empiricist views incorrect. Kepler insisted that a scientific hypothesis should have a rational basis. Hence the "genius" view (i.e., "There is no logic of discovery") does not hold good here. We have seen that even the so-called accidental discoveries were products of clear rational reasoning and analysis in at least one very good sense of "rationality": that the discoveries were made on the basis of considered principles. Indeed, these "accidental discoveries" emerged as a result of his strict adherence to certain principles of rationality.

At the same time, it is clear that Kepler's discovery could not have been achieved merely by a process of logical deduction from certain principles. Although, as we have seen, there is an identifiable Keplerian method of science, he did not give us any strict rules of logic which would yield his laws, as it were, automatically.

It follows, therefore, that none of the neat patterns or schemata ordinarily presented in the study of scientific discovery seems to fit Kepler. The H–D method, as it is traditionally understood, does not match Kepler's work exactly.[2] Peirce and Hanson argue that Kepler's process was a paradigm for the method of retroduction. From our study it is clear that Hanson's analysis is too quick and simplified to tell the whole or true story. He schematizes the retroductive method as follows: "1. Some surprising phenomenon P is observed. 2. P would be explicable as a matter of course if H were true. 3. Hence there is reason to think that H is true."[3] I think that the crucial step in this schema is the second proposition. How does one come to realize that P would be explained as a matter of course if H were true? Again, from where does H come? In the absence of adequate answers to these questions, the retroductive method is not of much help in the analysis of scientific discovery. Hanson describes the case as follows: "Kepler's task was: given Tycho's data, what is the simplest curve which includes them all?"[4] Kepler struggled back and forth with these data, juggled various hypotheses and at last came up with the elliptical orbit. This process seems to be a "curve-fitting" operation in disguise, except that here the curves are ready-made. For basically what Hanson seems to argue for is that Kepler had the data and some ideas about certain curves. In his work he tried the various options one after the other, and finally he fell on the ellipse. Hanson is surprisingly silent about the source of these various hypotheses. Why did Kepler choose these options rather than some other ones? Hanson seems to imply that the choice was governed solely by empirical considerations since he does not point out any other factors in this context. If his view were correct, then Tycho should have been the first to discover the correct path. Again, Fabricius should have made the discovery without any difficulty, because all these data were available to him and he was also seriously interested in the study of Mars. In fact, Fabricius himself tried to fit observational data to different curves, but gave up the project in despair. Hanson's view cannot explain satisfactorily why Kepler arrived at several crucial points of discovery, e.g., discovery of the constancy of inclination of planetary planes and the "rediscovery" of the equant, long before he had a chance to study Tycho's data. All these considerations go to show that the retroductive schema Hanson proposes is too watered-down to be of much help to give a clear and accurate understanding of Kepler's process of discovery. Hanson's proposal fails to provide any clear way for taking into account the role of the extra-empirical factors. Hence the attempt to fit Kepler's discovery straight into the logic of retroduction is highly unsatisfactory.

These considerations show that, despite the fact that for so many philosophers of scientific discovery Kepler's case has been a gold mine for developing their theories, none of the so-called standard patterns seems to suit his process of discovery. Indeed, some specific features of these patterns fit his work well, but not all of them. In this way our findings on Kepler's work seem to discourage anyone searching for a neat, logical mechanism or algorithm to account for scientific discoveries. Hence there seems to be an element of unpredictability or unprogrammability about scientific discovery, at least as far as Kepler's work goes.

However, we can identify some of the elements which marked him from most of his contemporaries and hence can be looked upon as possible explanations for his creativity and originality. In Kepler we see an inner freedom in the sense that, although he took his religious, philosophical, and empirical principles seriously, he was never inseparably or passionately attached to them. He did not consider any principle inviolable. The paradigm in point is his final rejection of the circularity principle, which even Galileo considered to be inviolable. Rather, Kepler emphasized a particular principle at one time and another at other times; sometimes he was willing to compromise one principle, at other times he would tolerate no such compromise. He was willing to accept that in one case a particular principle (or combination of them) took precedence over another (or several others). These considerations were not arbitrary, however; they were backed by reasons widely considered to be valid at that time.

This flexibility was not confined to Kepler's philosophical principles only; in his scientific positions also we see a similar attitude. For instance, although in the beginning he subscribed to an immobile sun, in the light of further research he unhesitatingly accepted a rotating sun.[5]

Closely associated with the above idea is his willingness to challenge any idea, his refusal to accept any idea merely on authority. The critical spirit was the hallmark of his scientific work.

An active and fertile imagination was another special characteristic of Kepler. He gave almost free rein to his imagination. However, he subjected these ideas to critical test and evaluation in the light of his principles, accepting only those which could pass the test successfully. His work teaches us that success comes, not by having only the right ideas and by making only the right moves, but in having many ideas, some false and some true, and subjecting them all to strict test and evaluation. In this sense he seems to have anticipated the basic "falsificationist spirit" of Karl Popper.

The transitional character of Kepler, beautifully expressed by Hanson and others, is brought out more fully and explicitly in this work. I have tried to identify both the traditional and the original, or new, elements in his work, and have analyzed how these new ideas came about and how they contributed to the creation of a new science.

With regard to the issue of the interaction between science and religion, Kepler's case makes a remarkable contribution. It asserts that the two need not be competitors but can be collaborators, helping each other towards mutual enrichment and development. One can simultaneously be both a scientist and a religious believer. At the same time, this study shows that such an integration of science and religion can be achieved only if religious and scientific positions allow a certain flexibility. Most of the religious leaders like Hafenreffer remained inflexible and so could not go along with Kepler. Similarly, most of the scientists of his day, like Fabricius, were unwilling to introduce changes in their scientific ideas and thereby failed to be with Kepler. Both Hafenreffer and Fabricius subscribed to a compartmentalist view, the view well expressed in sayings such as "When Faraday opens the door of his laboratory, he closes the door of the oratory." Kepler, on the other hand, believed that a scientist could keep his laboratory open without closing his oratory.

This book, therefore, presents a vastly more adequate picture of Kepler's substantive thought. Although it attempts to remedy some of the limitations in Kepler scholarship, pointed out in the introduction, it has a number of limitations. We have seen that in Kepler's discovery the interaction, or working together, of all three factors was required. Can such an interaction be considered normative for all science? Although I emphasize the positive and constructive role of philosophy and religion in his scientific work, I do not propose their use as normative for all good science. I do not argue that they are, much less that they must be, always present in scientific reasoning. That more general question about the "logic of discovery" is a subject for future work. But we at least see here an important case in the history of science in which philosophical and religious ideas did play a central role. Hence any adequate philosophical interpretation of scientific change must allow for the role of such principles in the development of science: it must be recognized that such principles can play a legitimate part in that development, whether or not they are operative throughout.

A reader may point out that in the present study, I have discussed only how the philosophical and the religious influenced Kepler's scientific work, and have left out the reverse case. For instance, how did Kepler's acceptance of Copernicanism influence his religious views? Certainly, his perception of the new view as simple, ordered, harmonious, and, above all, true must have strengthened his idea of God as the most simple and the most true. Although I will discuss shortly one of the major consequences of the influence of his science on his religion, a detailed study of this influence in each specific step will necessarily take us to an elaborate investigation of his various theological writings. Since such a research is beyond the scope of this present study, I leave it also as a subject for future work.

This new and richer interpretation of Kepler's work and his methods of thinking offers the possibility of a correspondingly richer and more adequate assessment of his importance as a thinker—not only in being a major contributor to the subsequent development of science, but also of theology and philosophy. Such a new assessment reveals that his influence and importance were all pervasive, in the sense that they brought about a transformation in all the three major areas: religion or theology, philosophy, and science.

A NEW ASSESSMENT OF KEPLER'S INFLUENCE

Influence on Theology:
The Birth of a New Outlook towards
Scriptural Interpretation

The influence of Kepler's scientific ideas on his religion was most conspicuous in his new outlook on scriptural interpretation. The specific problem confronting Kepler was how to reconcile astronomy, specifically the Copernican astronomy, with the Bible. He was caught up in a difficult dilemma: on the one hand, he was convinced that the Copernican system was true; on the other hand, he had no doubt about the veracity of the Bible. But there are many passages in the

Bible which seem to contradict the Copernican position. How can such a contradiction arise? How can one truth contradict another on the same issue? Can a rational God allow such a situation? Kepler put the conflict in another way also. God has both tongue and hands (fingers). The Holy Book is the word of God, the work of God's tongue. The Book of Nature is the deed of God, the work of God's hands. Insofar as astronomy is the study of this Book of Nature, true astronomical laws are nothing but laws governing the works of God's hands. To say that astronomy and the Holy Scripture contradict each other is tantamount to denying any coordination between God's tongue and hands. Since such can never be the case, he concluded: "Therefore in matters which are quite plain every one with strong religious scruples will take the greatest care not to twist the tongue of God so that it refutes the finger of God in nature."[6] Obviously, Kepler's dilemma was to find a rational way to bring about a reconciliation between two truths deepest to his heart: the veracity both of the Bible and of the Copernican system.

Basically, Kepler's solution was that there cannot be a real conflict between scientific truth and scriptural truth. The contradiction, only apparent, not real, arises because of a literal interpretation of Holy Scripture. He advocated a nonliteral interpretation, especially in passages where Scripture talks about scientific issues. In this way his scientific views and his faith in the ability of science to reveal truths about nature moved him to argue for a nonliteral interpretation of Scripture.

The official reaction of the churches towards the problem ranged from outright condemnation of the Copernican position to tactful conditional tolerance of it. Luther, for instance, condemned Copernicanism as early as 1539.[7] Cardinal Bellarmine, despite all the adverse publicity he received because of the role he was commissioned to play in the Galilean episode, was very reasonable. He was ready to respect the Copernican position under one condition, which he believed could not be satisfied. In 1615 he wrote to the Copernican, Foscarini: "If there were a real proof that the Sun is in the center of the universe, that the earth is the third heaven, and that the sun does not go around the earth but the earth round the sun, then we should have to proceed with a great circumspection in explaining passages of Scripture which appear to teach the contrary, and rather admit that we did not understand them than declare an option to be false which is proved to be true."[8] After giving this general condition he added: "But as for myself, I shall not believe that there are such proofs until they are shown to me."[9]

The church authorities of the day, in general, subscribed to a compartmentalist view concerning science and religion. According to them, the introduction of scientific findings in the interpretation of Scripture was trespassing into sacred territories. Already on April 12, 1598, Hafenreffer sternly reminded Kepler:

> Therefore, if there is any place for my brotherly advice (as I firmly hope), in demonstrating such hypotheses you will play the part of the pure mathematician, without worrying at all whether the hypotheses do or do not conform to the things of creation. For the mathematician accomplishes his purpose, I believe, if he produces hypotheses to which the phenomena correspond as closely as possible. And you yourself, I suppose, will yield to the man who could discover better hypotheses.[10]

Kepler's former teacher was undoubtedly voicing the predominant view in the theological circles of the day.

Kepler's firm faith in science and in the principle of unity of nature could not accept such a sharp dichotomy. Although he always remained respectful towards the word of God, he argued that it had to be interpreted in the light of scientific findings. In this context he subscribed to the accommodation theory of interpretation of the Bible, according to which God in revealing to humans accommodates language and expression to the people God reveals to and to the purpose God has in mind. As Kepler wrote to Herwart, the inspired writers of the Bible used the ordinary language and concepts of the day to communicate God's message, "not for linguistic accuracy, but for the sake of conversing and communicating his message."[11] The accommodation theory argues that we must recognize that a real distinction exists between *what* is said and *how* it is said. On the other hand, a literal interpretation denies such a distinction and identifies what is said with how it is said. Kepler's position was well founded on the basic ideas of his religion. God the creator respects humans created in God's own image. God respects also the conditions and the historical situation humans are in when revealing to them. Again, because supremely rational, God would follow the rational strategy of subordinating means to goal. God would adjust methods of communication in such a way that the message would get across to people most effectively.

In opposition to many Lutheran theologians of his day who looked upon the Bible as a textbook of astronomy, Kepler argued that the purpose of the Holy Book is not to teach astronomy, but moral conduct. According to him, except for the first chapter of Genesis, the Holy Writ is not meant to instruct humans in natural science.[12] His view was basically the same as what Cardinal Baronius said: "The Bible teaches us how to go to heaven and not how the heavens go."[13] The Bible wants to take the believers to a higher level of morality, not to the level of the study of the planets and stars. "For astronomy discloses the causes of natural phenomena and takes within its purview the investigation of optical illusions. Much loftier subjects are treated by Holy Writ. . . ."[14] As an illustration, Kepler referred to the Book of Job, chapter 38, where God talks of the creation of the world. Although this passage refers to topics ordinarily dealt with in astronomy, it is not an astronomical (scientific) analysis on how the earth and things in and around it have been formed. The purpose of this passage is to remind humans of the supremacy of God over all creation, to remind them of their ignorance, and to recall them to humble submission to and reverence for God, the Master of all creation.

How about the passages in the Bible often quoted by the opponents of Copernicus? Kepler argued that the principles developed above could show that these passages did not contradict Copernicanism. The most frequently quoted one is from Joshua 10:12–13: "Sun, stand thou still at Gibeon, and thou Moon in the valley of Aijalon. And the Sun stood still, and the Moon stayed, until the nation took vengeance on their enemies." The usual argument (in Kepler's day) had been that if the sun were stationary, it was pointless for God to order it to stop moving in answer to Joshua's prayer. According to Kepler, the leader of the Israelites was not speaking as an astronomer, he was using the language of the people. God, on the other hand, understood what Joshua wanted and granted it. "The sum of Joshua's prayer came down to this that it might so appear to him, regardless of the reality; to be sure, the appearance was not groundless and invalid but was related to the desired effect."[15] He continued: "Thoughtless people consider it only a contradiction of words: 'the Sun stood still,' and 'the Earth stood still.' They do not consider that this contradiction arose only within the boundaries of optics and astronomy, and for this reason it does not extend to common usage."[16] Definitely in astronomy it is important to know which heavenly body stood still, but in ordinary conversations such distinctions are immaterial. The relevant question is What does the author intend? The conclusion Kepler wanted to draw is clear: if the Bible is a book on astronomy and if Joshua was speaking as an astronomer, then there is a serious problem and the Joshua passage can be taken as a refutation of the Copernican system. But the conditionals are not true.

Kepler's arguments are quite persuasive. However, there are serious difficulties with his view. His position implies that some passages of the Bible are to be taken nonliterally, whereas some others literally. But how do we know which is to be taken nonliterally and which literally? Kepler would not want to say that all passages referring to scientific issues should belong to the nonliteral category because he himself gave the first chapter of the Book of Genesis as a clear exception to this rule. An absence of a clear criterion would render his view an easy prey to inconsistency. The main criterion at that time was articulated by Bellarmine in the passage quoted above: in order for a scientific theory to demand a nonliteral interpretation of a passage in the Bible, it is necessary that the scientific theory concerned be demonstratively[17] true. One must show that the scientific theory cannot be false. It is true that this was the Catholic position and Kepler did not have to abide by this. However, as far as I know, the other Christian denominations with which Kepler had close contact took an even more stringent position.[18] Hence I believe that Kepler accepted Bellarmine's criterion. This is clear from his statement right in the beginning of the first chapter of MC: "I promise generally that I shall say nothing which would be an affront to Holy Scripture, and that if Copernicus is convicted of anything along with me, I shall dismiss him as worthless."[19] Hence it was necessary for Kepler to be absolutely sure of the truth of Copernicanism before he could opt for a nonliteral interpretation of biblical passages which seemed to go against the new theory. He was convinced of the veracity of the new theory and so advocated the new interpretation of the Bible. Thus this important contribution of Kepler towards scriptural interpretation was a result of his deep faith in scientific view in general, and the Copernican view in particular.

Kepler's Influence on Philosophy

Several of the original ideas developed by Kepler had tremendous influence on later philosophical thinking. Since I have discussed many of these points in detail earlier, I shall only mention them more summarily here. The principle of sufficient reason was a great Keplerian contribution to philosophy: many philosophers, especially Leibniz, adopted this principle fruitfully in their work. For the development of the principle of precision and therefore of empiricism, in the sense that a theory is responsible for every detail of experience, and the principle of rationality—commonplace in modern thinking—we are also indebted to our astronomer. Further, Kepler's assertion that the universe is mathematical and his development of a mathematical view of the world molded later thinking in a significant way, especially in the development of modern mathematical physics; often Descartes is given credit for this, but it is clear from our investigation that Kepler's ideas predated those of Descartes. I have also pointed out how Kepler helped shift the medieval emphasis on formal and final causes to the modern emphasis on efficient cause.

Kepler's Influence on Science and the Philosophy of Science

THE BIRTH OF MECHANICAL PHILOSOPHY

Mechanical philosophy can be said to have cooriginated with modern science. It gradually developed in the seventeenth and eighteenth centuries and eventually became the official philosophy of science until its final collapse towards the end of the nineteenth century. Mechanical philosophy, as Shapere interprets it, holds that mechanical explanation is the best, if not the only true, explanation of material phenomena. It proposes to explain physical nature, if not nature in general, in terms of matter in motion and its interaction. More specifically, it postulates that there are four fundamental concepts by which every thing in the universe can be accounted for: matter (mass), space, time, and force (interaction). Traditionally, this philosophy is associated with the names of Galileo, Descartes, Leibniz, and Newton. However, an unbiased look at Kepler's contributions cannot but reveal that he, too, must be placed among these eminent scientists. In recent years, Hanson, Mittelstrass, and others have argued that Kepler is one of the founders of modern science and mechanical philosophy. According to Hanson, "the line between Ptolemy and Copernicus is unbroken. The line between Copernicus and Newton is discontinuous, welded only by the mighty innovations of Kepler."[20] From our study it is clear that Kepler with his original ideas on scientific explanation, force, mass, and scientific method made significant contributions to the origin and development of mechanical philosophy.

That Kepler strove to build up a world system on the basis of mechanical philosophy he made clear in a letter to Herwart: "My aim is to show that the heavenly machine is not a kind of divine, living being, but a kind of clockwork (and he who believes that a clock has a soul, attributes the maker's glory to the work), insofar as nearly all the manifold motions are caused by the most simple, magnetic, and material force, just as all motions of the clock are caused by a simple weight."[21] Koyré points out that, although the comparison between the so-called celestial machine and a clock was made by others like Buridan and Rheticus before him, Kepler was the first to postulate a definite identification of machine with clock.[22] There is a marked difference between comparison and identification. A comparison of the celestial world with a clock is quite compatible with treating the heavens as a living organism, since certain properties of a living organism can be compared with those of a clock, e.g., movement of component parts. This could explain why, although people made the comparison between a clock and the celestial machine, this move did not lend itself to the development of a mechanical philosophy. Kepler, on the other hand, identified the two, which was an important first step towards deanimating the celestial world, in the sense of replacing the idea of planetary souls with the idea of force and of developing a mechanical philosophy.

According to Kepler, an adequate scientific explanation should be given in terms of real, physical forces seated in matter, the demand mechanical philosophy also makes. Of course, as we have noted, for him scientific explanation should meet a further requirement of being consistent with certain accepted philosophical and religious principles. I have shown that he contributed significantly towards the development of modern ideas of force and mass. These new ideas also aided mechanical philosophy further along its growth.

THE BIRTH OF THE NEW SCIENCE

In a remarkable way, Kepler's empirical ideas transformed the existing science. Mittelstrass goes to the extent of saying that they effected not just a modification, but a "fundamental, namely a foundationally-determined change."[23] I shall argue that these changes affected the goal, the domain, the method, and the criteria of science.

New Goal of Science:
From "Saving the Phenomena" to Explaining the Phenomena

Following the Platonic principle, the old tradition of astronomy set its goal as "saving the phenomena." Thus the goal of astronomy was to account for observed phenomena in terms of circular orbits. It was not its purpose to go beyond the observed to investigate the underlying reality. In opposition to this view, Kepler insisted that astronomy should go beyond mere calculation and prediction of motion to investigate the nature of things. His view he made clear on several occasions, especially in the *Apologia*. Attacking Ursus's defense of the traditional position of astronomy, Kepler wrote:

> For even though what he [Ursus] mentions is the primary task of the astronomer, the astronomer ought not to be excluded from the community of philosophers who inquire into the nature of things. One who predicts as accurately as possible the movements and position of the stars performs the task of astronomy well. But one who, in addition to this, employs true opinion about the form of the universe performs it better and is held worthy of greater praise.[24]

To the Aristotelians who opposed such a move he had the following harsh words to say: "O wretched Aristotle! Wretched is that philosophy of his in the part said to be most excellent, if astronomers become such faithless witnesses on philosophical questions."[25] Kepler thus redefined the role of astronomy as not just saving the phenomena, but also explaining them.

New Domain of Science:
The Fusion of the Celestial and the Terrestrial

One of the principal tenets of the old worldview was that the celestial and the terrestrial were two separate and distinct worlds.

They were separate ontologically and epistemologically: ontologically because the two were made up of distinct and mutually irreducible elements; epistemologically because the two worlds could be known only through completely different means. For instance, as I have discussed in part 1, mathematics was banned from the study of the material terrestrial world, whereas it was a perfectly fit tool in the investigation of the celestial realm. Kepler's work attempted to demolish this separation in both these aspects and build up a unified cosmic view and a unified science. He was the first to do so seriously and in a constructive way.

One may argue that the impetus theorists like Buridan, Oresme, and other followers were attempting to carry out such a project two centuries before. True, the impetus theory of Buridan in some limited ways could be interpreted as a move in this direction. However, it was a far too limited project. It was proposed as a solution to the problems facing the Aristotelian theory of projectile motion. Thus the impetus theory was limited to a specific case; its aim was not to find a universal law applicable to both the celestial and the terrestrial. Not only that, it remained very much on the speculative level, lacking any experimental study or verification, though it appealed to occasional "thought experiments" and experience. Furthermore, it was not even advanced as truth by Buridan and Oresme, but as an argument to assert the inability of reason to prove anything.[26]

Kepler, on the other hand, searched for a single universal law governing both worlds. He unhesitatingly applied the laws of mechanics to the study of the motion of heavenly bodies. In this way he redefined the domain of physical science as comprising both the terrestrial and the celestial.

New Method of Science: The Fusion of Mathematics and Physics

As I have already pointed out, Kepler's scientific method had many novel elements. Perhaps the most significant among them was his insistence that mathematics be an integral part of the scientific study of natural phenomena. All through his work one can see a constructive and fruitful fusion of mathematics and physics. Today, mathematics is so closely associated with physics that it is practically impossible to make any progress in physics without it. We take the fusion of mathematics and physics for granted. But we have already seen that this was not the case in Kepler's day. Since the celestial and the terrestrial were epistemologically distinct from one another, the methods used for one could not be used for the other. Mathematics could be used in the celestial world but not in the terrestrial world, whereas methods of physics could be used in the terrestrial but not in the celestial.

Kepler refused to accept this dichotomy between mathematics and physics. Not only did he believe that the same physical laws were applicable to both worlds, but he also held that both were amenable to mathematical investigation. Indeed, he believed that our universe by its very nature is mathematical and hence the use of mathematics must be the key to unlocking the secrets of the cosmos. For him, the use of mathematics was not just an option, not just a convenient tool, but a necessity. Thus Kepler was the first mathematical physicist.[27]

New Criterion in Science: The Emphasis on the Physical

While discussing Kepler's idea of scientific explanation, I pointed out that he insisted on the physical aspect in any satisfactory scientific explanation. This stress on the physical expressed itself in different ways in Kepler's work. For him, physical explainability became a criterion of acceptability for any scientific hypothesis. If a scientific hypothesis or method lacked a physical foundation, it was neither true nor acceptable, e.g., epicycles. In the introduction to AN he wrote: "No method would succeed except that which enters from the actual physical causes of the motions, which I establish in this work."[28] Even if a hypothesis agreed with the phenomenon to be explained, he professed that he would not accept it unless it was physical. For instance, in a letter to Longomontanus he wrote: "I have a hypothesis which I formulated four years ago already which puts the planet for me on the eccentric in the right places with most scrupulous accuracy. But it doesn't please me since it isn't a physical hypothesis, but is what I call hypothesis."[29] This passage gives witness that Kepler made a sharp distinction between a physical hypothesis and a mere hypothetical construction. Whatever be the specific details of the hypothesis referred to in this letter, the point is clear: no scientific hypothesis can be adequate without being physical.

Physical explainability again became his criterion for choosing between two equipollent hypotheses. He preferred the physical to every other alternative, even if the physical could account for the phenomenon under consideration no better than its rival. Replying to Longomontanus, a staunch supporter of Tycho, Kepler wrote: "It is possible that someone else will be able to present the same by retaining the hypothesis of Tycho. But I have not been so fortunate. I profess I favor the physical form which I embrace in preference to others so long as it carries out its functions as well as the others."[30] Kepler gave an illustration of this position in one of his letters to Maestlin.[31] Hence in his new science physical explainability was considered an important criterion for the choice and acceptance of theories.

Thus Kepler gave a new definition and a new direction to astronomy. Since astronomy at that time was the paradigm of physical science, the birth of a new astronomy was actually the birth of a new science. A new era for science was born. Physical science in general and astronomy in particular could never be what they used to be. In recent times, many scholars have noted the significance of this contribution. This awareness has prompted Mittelstrass, for example, to talk of a "Keplerian revolution" rather than a "Copernican revolution." Hanson talks of a "Copernican disturbance" and a "Keplerian revolution." Perhaps these scholars are somewhat overreacting to the unjust treatment that has been meted out to Kepler by scholars until recently. At the same time, I believe that Kepler's contributions are unique and significant enough to earn for him a permanent place among the founders of modern science.

NOTES

INTRODUCTION

1. Recently some scholars have cast doubts on this claim. See Bruce Stephenson, *Kepler's Physical Astronomy* (New York: Springer-Verlag, 1987), p. 3. See also W. H. Donahue, "Kepler's Fabricated Figures: Covering Up the Mess in the *New Astronomy*," *Journal for the History of Astronomy* 19 (1988), 217–237.

2. For instance, commenting on the *De Motu* of Newton, I, Bernard Cohen writes: "Kepler's laws are not strictly true in the world of physics but are true only for a mathematical construct. . . . The distinction Newton draws between the realm of mathematics, in which Kepler's laws are truly laws, and the realm of physics, in which they are only 'hypotheses,' or approximations, is one of the revolutionary features of Newtonian celestial dynamics" ("The *Principia*, Universal Gravitation, and the 'Newtonian Style,' in Relation to the Newtonian Revolution in Science," in *Contemporary Newtonian Research*, ed. Zev Bechler [Dordrecht: Reidel, 1982], p. 46). See also Pat Langley, Herbert A. Simon, Gary L. Bradshaw, and Jan M. Zytkow, *Scientific Discovery* (Cambridge, Mass.: MIT Press, 1987), p. 27.

3. See Arthur Koestler, *The Watershed* (New York: Doubleday Anchor Books, 1960).

4. Robert Small and J. L. E. Dreyer, although they do not subscribe to such an extreme positivist view, are not too far from it. See

Robert Small, *An Account of the Astronomical Discoveries of Kepler,* reprint of the 1804 text (Madison: University of Wisconsin Press, 1963). See also J. L. E. Dreyer, *A History of Astronomy from Thales to Kepler,* 2d ed. (New York: Dover, 1953), pp. 372–424. Robert Westman also seems to lean towards this view, though in a much less emphatic way, as I shall point out in part 2. See his "Johannes Kepler's Adoption of the Copernican Hypothesis" (Ph.D. Dissertation, University of Michigan, 1971). In this connection see also D. T. Whiteside, "Keplerian Planetary Eggs, Laid and Unlaid, 1600–1605," *Journal for the History of Astronomy* 5 (1974), 1–21.

5. For instance, see Curtis Wilson, "Kepler's Ellipse and Area Rule—Their Derivation from Fact and Conjecture, in *Kepler: Four Hundred Years: Proceedings of Conference Held in Honor of Johannes Kepler,* ed. Arthur Beer and Peter Beer, *Vistas in Astronomy* 18 (New York: Pergamon, 1975), p. 587; henceforth this volume will be referred to as BB.

6. Norwood Russell Hanson, *Patterns of Discovery* (Cambridge: Cambridge University Press, 1961), p. 85.

7. See Edward Strong, *Procedures and Metaphysics* (Hildesheim: Georg Olms, 1966). Among those who discuss the influence of philosophical and religious principles, Edward Strong is the most emphatic in denying any role to these extrascientific or nonempirical factors in the discovery of Kepler's laws. Strong attempts to present arguments calculated to show that philosophy and religion were not needed for Kepler's discovery. Accordingly, philosophy or metaphysics had no fundamental role to play in the origin and development of science. "Mechanical knowledge [i.e., scientific knowledge] marches by method, not by metaphysics" (p. 8). Strong argues that neither in Galileo's nor in Kepler's work do we see metaphysics having any decisive part. In this view, apart from supplying some additional confidence to the scientist, metaphysical considerations have no real role in scientific discovery (see p. 169). Not only does Strong ban metaphysics from scientific discovery, he banishes religion as well. In Kepler's scientific work, religious beliefs were not fundamentally functional, and, in fact, religion hindered his becoming a true scientist: only when Kepler rid his work of such factors could he become a true scientist (see p. 167). My own investigations lead me to conclude that Strong's positon does not square with the facts.

8. See Curtis Wilson, "How Did Kepler Discover His First Two Laws?" *Scientific American* 93, 226 (March, 1972), 93. Eric Aiton, like Wilson, believes that more than empirical data was involved in the discovery of the laws (see his "The Elliptical Orbit and the Area Law," BB, p. 573.) Owen Gingerich is another contemporary scholar

willing to assign a similarly limited role to nonempirical factors (see his "Kepler," in *Dictionary of Scientific Biography*, vol. 7 [New York: Charles Scribner's, 1973], p. 307).

9. Alexandre Koyré, *The Astronomical Revolution*, tr. R. E. W. Maddison (Ithaca, New York: Cornell University Press, 1973), p. 120.

10. For instance, see Edwin Burtt, *The Metaphysical Foundations of Modern Science* (New York: Doubleday Anchor Books, 1954), pp. 36–71. In his book already cited, Koestler also follows a similar trend.

11. Gerald Holton, "Johannes Kepler's Universe: Its Physics and Metaphysics," *American Journal of Physics* 24 (1956), 351.

12. Ibid., p. 340. Notice that the confusion is only apparent. In reality there was no confusion since, as I will show, incongruous elements like physics and metaphysics collaborated in Kepler's discovery process to produce positive results.

13. Ibid., p. 351.

14. Gerald Holton, "Johannes Kepler: A Case Study on the Interaction between Science, Metaphysics, and Theology," *Philosophical Forum* 21 (1956), 21. Although this essay has a different title, it is basically the same as the one I quoted earlier, at least as far as the question of Kepler's discovery of the laws is concerned.

15. A number of important studies on these aspects have been published recently. For instance, for Kepler's religious views see Jürgen Hübner, *Die Theologie Johannes Keplers zwischen Orthodoxie und Naturwissenschaft* (Tübingen, 1975). For Kepler's ideas on mathematics, see Judith Field, *Kepler's Geometrical Cosmology* (Chicago: University of Chicago Press, 1988). For his scientific ideas, see Bruce Stephenson, *Kepler's Physical Astronomy* (New York: Springer-Verlag, 1987).

16. Letter on March 5, 1605, in *Gesammelte Werke*, ed. W. von Dyck, Max Casper, F. Hammer, M. List (Munich, 1937–) [hereafter GW], vol. XV, nr. 335: ll. 121–122. The announcement to Fabricius was on Oct 11, 1605.

17. Letter on March 28, 1605, in GW XV, nr. 340: ll. 320–324.

18. For instance, see Kepler's *Tertius Interveniens*, Section 126. This was published in 1610, i.e., soon after AN. Similar ideas can be found in his *Dissertatio cum Nuncio Sidereo*, also published in 1610. Both MC and HM emphatically proclaim the centrality of God and religion in all his thinking and work. The foundational significance of MC for all his subsequent astronomical work Kepler declared in the "Dedicatory Epistle" for the 2d editon, published in 1621. He wrote: "Almost every chapter on astronomy which I have published since that time [i.e., 1596, the year of publication of the 1st edition] could be referred to one or another of the important chapters set out in this little

book, and would contain either an illustration or a completion of it"
(MC, p. 39). Obviously, religious principles were an integral part of his
system all through his career.

1
KEPLER'S RELIGIOUS IDEAS

1. Johannes Kepler, *Mysterium Cosmographicum*, tr. A. M. Duncan (New York: Abaris Books, 1981), p. 53.

2. Kepler's letter to Fabricius, dated July 4, 1603, in GW XIV, nr. 262: ll. 495–496.

3. Amos Funkenstein, in his recent, thought-provoking book discusses the conceptual changes that came about in certain theological themes as a result of the scientific revolution of the sixteenth and seventeenth centuries. In particular, he focuses on the omnipresence, the omnipotence, and the providence of God. Although he investigates in detail the contributions of Descartes, Galileo, Leibniz, Newton, and others towards these developments, very little is said about Kepler's role (see *Theology and the Scientific Imagination from the Middle Ages to the Seventeenth Century* [Princeton: Princeton University Press, 1986]). Our discussion in this and the subsequent chapters will show that Kepler also made a substantial contribution towards these conceptual changes.

4. It may be noted that Jainism and Hinayana Buddhism deny the existence of any supreme being. Hence the description given here applies only to religion in general, not to all recognized ones.

5. Max Caspar, *Johannes Kepler: 1571–1630*, tr. Doris Hellman (New York: Abelard-Schuman, 1959), p. 374.

6. The polyhedral theory explains the arrangement of the different planets in the solar system in terms of the five regular solids that are nested within a series of concentric spheres. Kepler found that by placing appropriately the five regular solids within concentric spheres, one could closely approximate the spacing of the planets.

7. Letter to Maestlin, dated August 2, 1595, in GW XIII, nr. 21: ll. 60–63.

8. See MC, p. 63.

9. MC, p. 225.

10. GW III, p. 33: ll. 10–16, tr. O. Gingerich and W. Donahue, "*Astronomia Nova*," in *Great Ideas Today*, ed. Mortimer Adler and John van Doren (Chicago: Encyclopedia Britannica, 1983), pp. 321–322.

11. GW VI, p. 289: ll. 7–10, tr. Rudolf Haase, in "Kepler's Harmonies, between *Pansophia* and *Mathesis Universalis*," BB, p. 524.

12. GW VI, p. 287, tr. Charles Glenn Wallis, in *Ptolemy, Copernicus, Kepler*, Great Books of the Western World 16, ed. Mortimer J. Adler (Chicago: Encyclopedia Britannica, 1982), p. 1009. Henceforth this item will be referred to as GB.

13. GW VI, p. 290: ll. 3–6, tr. GB, p. 1010.

14. Ibid., p. 368: ll. 14–22.

15. See Claus Schedl, "Die logotechnische Struktur des Psalms VIII und Keplers Weltharmonik," in *Johannes Kepler: 1571–1630, Gedenkschrift der Universität Graz* (Graz, 1975), pp. 105–123. See also Jürgen Hübner, "Naturwissenschaft als Logpreis des Schöpfers," in *Internationales Kepler-Symposium Weil der Stadt, 1971* (Hildesheim, 1973), pp. 335–356.

16. Kepler's letter to Mathias Bernegger, dated October 21, 1630, in GW XVIII, nr. 1145: ll. 34–35. See also *Johannes Kepler: Life and Letters*, ed. Carola Baumgardt (New York: Philosophical Library, 1951), p. 192. Henceforth Baumgardt's book will be referred to as CB.

17. MC, p. 107.

18. Gerald Holton, "Johannes Kepler's Universe: Its Physics and Metaphysics," p. 350.

19. MC, p. 29.

20. GW VI, p. 354: ll. 5–6.

21. MC, p. 93.

22. Jürgen Hübner, *Die Theologie Johannis Keplers zwischen Orthodoxie und Naturwissenschaft* (Tübingen, 1975), p. 191.

23. See GW XIII, nr. 23: ll. 78–84.

24. See MC, pp. 63 and 93.

25. See GW II, p. 19: ll. 10–22.

26. See GW XV, nr. 340: ll. 320–324.

27. See GW XVI, nr. 493: ll. 1–3.

28. See GW IV, p. 246: ll. 9–22.

29. See GW VII, p. 51: ll. 1–22.

30. See GW VI, p. 441: ll. 10–13.

31. Funkenstein argues that the works of many seventeenth-century scientist-theologians were marked by an attempt to render God more and more transparent (*Theology and the Scientific Imagination*, p. 25). Kepler's explanation of the Trinity-sphere seems to point along this direction. Later on we shall discuss Kepler's attempt to make another important Christian doctrine—that the human being is created in the image and likeness of God—more intelligible.

32. There are many forms and versions of voluntarism. I am describing very briefly only one of them.

33. Richard Taylor, "Voluntarism," *Encyclopedia of Philosophy*, vol. 8 (London: Macmillan, 1967), p. 271.

34. Kristian Moesgaard, "Copernican Influence of Tycho Brahe," in *The Reception of Copernicus' Heliocentric Theory*, ed. Jerzy Dobrzycki (Dordrecht: Reidel, 1972), p. 53.

35. GW XIII, nr. 21: l. 9.

36. See Kepler's letter to Fabricius, dated February 7, 1603, in GW XIV, nr. 248.

37. MC, p. 97.

38. GW VI, p. 223: ll. 32–33, tr. Wolfgang Pauli, "The Influence of Archetypal Ideas in the Scientific Theories of Kepler," in *Interpretation of Nature and of the Psyche* (New York: Pantheon, 1955), p. 166.

39. MC, p. 73.

40. GW VI, p. 271: ll. 30–31.

41. For instance, Christ in the gospels referred to himself as the light and the life.

42. GW VI, p. 364: ll. 26–28.

43. GW VI, pp. ll. 220–221, Kepler's marginal notes.

44. See Kepler's letter to Brengger in 1608, in GW XVI, nr. 488: ll. 390–398.

45. See A. H. Armstrong, *Plotinus* (London: George Allen and Unwin, 1953), pp. 69–71. In this context it is also remarkable that the Neoplatonists often identified the One with the sun.

46. See Proclus, *The Elements of Theology*, tr. E. R. Dodds, 2d ed. (Oxford: Clarendon Press, 1963), pp. 29–43. We shall take up these ideas in some detail in chapter 2.

47. GW VI, p. 219: ll. 23–24.

48. GW VI, p. 223: ll. 33–34, tr. Pauli, "Influence of Archetypal Ideas," p. 166.

49. Kepler's letter to Johann Georg Herwart von Hohenburg, dated April 10, 1599, in GW XIII, nr. 117: ll. 176–177, tr. L. H. Richardson, in *The Sidereal Messenger* 6 (1887), 137. Henceforth this translation will be referred to as SM.

50. GW XIII, nr. 117: ll. 174–176.

51. Letter dated September 14, 1599, in GW XIV, nr. 134: ll. 438–441.

52. Letter to Herwart, dated April 10, 1599, in GW XIII, nr. 117: ll. 177–179, tr. SM, p. 137.

53. Ibid., nr. 117: ll. 179–183.

54. Scholars like Funkenstein point out that one of the achievements of the scientific revolution was that it helped to narrow the gap

between divine and human knowledge. From this discussion one can surmise that in this respect also Kepler made a significant contribution.

55. GW VI, p. 271: ll. 30–33.

56. See GW VI, pp. 220–221, Kepler's marginal notes.

57. See letter, dated April 5, 1608, in GW XVI, nr. 488: ll. 396–401.

58. CB, p. 32.

59. Here one can see clearly the influence of Pythagoreanism and Neoplatonism. The former believed that engaging in the study of nature was a sure way to attain salvation, while the latter looked upon the universe as the manifestation of the Supreme Being.

60. See letter to Fabricius, dated July 4, 1603, in GW XIV, nr. 262: ll. 495–496.

61. GW XIII, nr. 117: ll. 295–296.

62. GW IV, p. 246: ll. 23–24.

63. Kepler's letter to Maestlin, dated October 3, 1595, in GW XIII, nr. 23: ll. 72–74.

64. Ibid., nr. 23: ll. 78–84.

65. MC, p. 71.

66. GW VI, p. 224: ll. 14–18, tr. Pauli, "Influence of Archetypal Ideas," p. 160.

67. Or it constitutes at least part of the essence of God.

68. *De Revolutionibus*, book I, ch. 10, reprint of the 1st edition of 1543 (New York: Johnson Reprint, 1965), tr. Edward Rosen, *On the Revolutions*, ed. Jerzy Dobrzycki (Baltimore: The Johns Hopkins University Press, 1978), p. 22. Henceforth Copernicus's original will be referred to as DR and Rosen's translation as ER. See also Hanson, *Constellations and Conjectures*, ed. W. C. Humphreys (Dordrecht: Reidel, 1973), p. 190.

69. GW IV, p. 308: ll. 36–37.

70. He wrote: "Consequently the sun has a far better claim to such noble epithets as heart of the universe, king, emperor of the stars, visible God and so on" (MC, p. 201).

71. GW VI, p. 363: l. 34, tr. GB, p. 1081.

72. Ibid., p. 364: ll. 1–5, tr. GB, p. 1081.

73. See GW VI, p. 364: l. 24–p. 365: l. 25.

74. See GW III, p. 35: ll. 5–17.

75. See Kepler's letter to Maestlin, dated October 3, 1595, in GW XIII, nr. 23: ll. 78–86.

76. MC, p. 107.

77. MC, pp. 53–55.

78. See Caspar, *Johannes Kepler*, p. 377.

79. This concept of harmony as deducibility of many phenomena

from a fundamental postulate is not the only sense in which it occurs in his works. He used it in at least one other sense. I will discuss it in chapter 2.

80. DR, book I, ch. 10, tr. Thomas Kuhn, *The Copernican Revolution* (New York: Vintage Books, 1955), p. 180.

81. Kepler's letter to Maestlin, dated October 3, 1595, in GW XIII, nr. 23: ll. 67–69.

82. "God, by certain selection in creating the world has, so to speak, taken such relationships out of himself. He has made order out of chaos, given form to matter, in accordance with the word of the Bible that everything is regulated by number, size, and weight" (Caspar, *Johannes Kepler*, p. 378).

83. Pistorius was the Father Confessor and adviser to Emperor Rudolf. Trained as a medical doctor, he was called Polyhistor by Kepler because of his encyclopedic knowledge.

84. See Pistorius's letter to Kepler, written on March 14, 1607, in GW XV, nr. 413: ll. 1–28.

85. Kepler's letter to Pistorius, dated June 15, 1607, in GW XV, nr. 431: ll. 21–24, tr. M. W. Burke-Gaffney, *Kepler and the Jesuits* (Milwaukee: Bruce, 1944), p. 39.

86. See Caspar, *Johannes Kepler*, p. 164.

87. This view has changed considerably, as is evident from contemporary writings in spirituality and theology.

88. GW VII, p. 25: ll. 29–31.

89. Kepler's letter to Maestlin, written on October 3, 1595, in GW XIII, nr. 23: l. 254.

90. Long before Kepler it was known to some of the Neoplatonist scholars.

91. GW III, p. 108: l. 3.

92. Quoted by Gingerich, in *Great Ideas Today*, p. 139.

93. ER, p. 7.

94. Kepler's letter to Herwart, written on March 25, 1598, in GW XIII, nr. 91: ll. 182–184. See Nicholas Jardin, *The Birth of History and Philosophy of Science: Kepler's "A Defense of Tycho against Ursus"* (Cambridge: Cambridge University Press, 1984), p. 9.

95. GW VII, p. 9: l. 12.

96. Even Tycho Brahe had a similar view. He argued that in theology and astrology it is inappropriate to ask for reasons. See Caspar, *Johannes Kepler*, p. 152.

97. See Thomas Aquinas, "Reason and Faith," in *Summa Contra Gentiles*, tr. English Dominican Fathers, reprinted in *Problems and Perspectives in the Philosophy of Religion*, ed. G. I. Mavrodes and S. C.

Hackett (Boston: Allyn and Bacon, 1967), p. 12.

98. Kepler's letter to Herwart, written on April 10, 1599, in GW XIII, nr. 117: ll. 173–174, tr. SM, p. 137.

99. See Caspar, *Johannes Kepler*, pp. 40–41.

100. A doctrine proposed by Luther, which argues that Christ is omnipresent at every enactment of the Lord's Supper. Funkenstein explains this doctrine as follows: "Christ's body was, and always is, permeated through and through by his divine nature. And, like God's power and essence, even Christ's body is everywhere at all times. The communion is only the occasion at which Christians are instructed by the word of God where to concentrate on finding Christ's presence" (*Theology and the Scientific Imagination*, p. 71).

101. GW XVII, nr. 750: ll. 245–247, tr. CB, pp. 106–107. I have modified the translation.

2
KEPLER'S PHILOSOPHICAL IDEAS

1. Caspar, *Johannes Kepler*, p. 60.

2. It may be noted that Kepler used the term "philosophical" in the broad sense it had in the seventeenth century. Hence his idea of philosophy included both the philosophical and the scientific. My point is that Kepler was both a philosopher and a scientist, in the modern sense.

3. GW III, p. 108: ll. 9–10, tr. in *Great Ideas Today*, p. 324.

4. Kepler's letter to Vincenz Bianchi, written on February 17, 1619, GW XVII, nr. 827: ll. 249–251.

5. Kepler's letter to Baron von Herberstein and the Estates of Styria on May 15, 1596, quoted and translated in CB, pp. 34–35. See also MC, p. 55.

6. It is obvious that in the traditional sense these principles involve the metaphysical and epistemological aspects.

7. See GW VI, pp. 364: l. 22–367: l. 40.

8. Plotinus, *The Enneads* V.I.6, tr. A. H. Armstrong (London: George Allen and Unwin, 1953), p.70.

9. This enables the emanation theory to get around the objection of pantheism.

10. Proclus, *The Elements of Theology*, tr. E. R. Dodds, 2d edition (Oxford: Clarendon Press, 1963), p. 25.

11. Ibid, Prop. 31, p. 35.

12. *Enneads* II.9.8, p. 106.

13. Pauli, "Influence of Archetypal Ideas," p. 197.

14. *Apologia pro Opere Harmonices Mundi*, GW VI, p. 446: ll. 22–25, tr. Pauli, "Influence of Archetypal Ideas," p. 200.

15. GW XV, nr. 305: ll. 18–20.

16. See Kepler's letter to Fabricius, written on December 2, 1602, in GW XIV, nr. 239: ll. 284–300. See also MC, p. 119.

17. "The *aspectus* are angular separations of the planets on the celestial sphere corresponding to certain fractional parts of the circle" (Aiton, MC, p. 240). The ancients had five such angular separations: conjunction (0 degree), opposition (180 degrees), quartile (90 degrees), trine (120 degrees), sextile (60 degrees). Kepler added three more: quintile (72 degrees), biquintile (144 degrees) and sesquiquadrature (135 degrees). The orientations of heavenly bodies can be such that they give rise to these *aspectus*.

18. See Pauli, "Influence of Archetypal Ideas," p. 189.

19. Idem.

20. See Kepler's letter to Galileo, written on October 13, 1597, in GW XIII, nr. 76: ll. 15–16. In fact, in this letter Kepler referred to Plato and Pythagoras as "our true masters."

21. Concerning the meaning and origin of the principle, "Save the phenomena," see Fritz Krafft, "The New Celestial Physics of Johannes Kepler," in *Physics, Cosmology, and Astronomy, 1300–1700: Tension and Accommodation*, ed. Sabetai Unguru (Dordrecht: Kluwer, 1991), p. 188.

22. Hafenreffer: "Ubi logos, ibidem ejus est caro." Kepler: "Ubi caro, ibidem est ó logos." Quoted by Hübner, *Theologie Keplers*, p. 203. See also Hübner, "Kepler's Praise of the Creator," BB, p. 371f.

23. Edward Grant, "Late Medieval Thought, Copernicus, and the Scientific Revolution," *Journal of the History of Ideas* 23 (1962), 212.

24. Here a priori method is understood in the usual sense. I will point out in chapter 3 that Kepler understood this method in a different way also.

25. Since mathematics has become an integral part of modern science, one may question the propriety of placing this principle in the section on philosophy. It may be noted that here we are considering mathematics as one of Kepler's general principles, principles that helped to form his worldview. Moreover, mathematics, as understood by him, involved entities not amenable to an empirical analysis.

26. GW VI, p. 223: ll. 34–35.

27. *De Stella Nova*, GW I, p. 192: l. 22.

28. Johannes Kepler, *Joannis Kepleri astronomi opera omnia*, ed. Christian Frisch, vol. 8 (Frankfurt, 1857–1871), p. 148. Quoted by E. A.

Burtt, *Metaphysical Foundations*, p. 68.

29. Written on April 9, 1597, in GW XIII, nr. 64: ll. 10–16.

30. Curtis Wilson, "Kepler's Ellipse and Area Rule—Their Derivation from Fact and Conjecture," BB, p. 590.

31. *Apologia*, GW VI, p. 397: l. 37.

32. See Stillman Drake, "Kepler and Galileo," BB, p. 239. Drake argues that this interpretation is based on a misrepresentation of Aristotle. In Drake's view the Philosopher advocated only a mild caution rather than a prohibition in the use of mathematics. For in the *Metaphysics* (955a, 15–17) he wrote: "The minute accuracy of mathematics is not to be demanded in all cases, but only in the case of things which have no matter. Hence its method is not the method of natural science, for presumably the whole nature has matter." Drake argues that this statement does not imply a prohibition. One may say that the Aristotelian view arose not because it was wrong or illegitimate to use mathematics in physics, but since the study of material nature involved matter, it could not yield mathematical accuracy. Hence mathematics would be useless in this study. However, saying that mathematics is useless for this purpose is not much different from asking not to use it. Furthermore, Aristotle did say, "Mathematical method is not the method of natural science," which hardly differs from a clear prohibition.

33. Quoted by Hübner, *Theologie Keplers*, p. 171.

34. See Kepler's letter to Herwart, written on July 12, 1600, in GW XIV, nr. 168: ll. 120ff.

35. See Kepler's letter to Maestlin, written on October 3, 1595, in GW XIII, nr. 23: ll. 54–55.

36. *Three Copernican Treatises*, ed. Edward Rosen (New York: Octagon Books, 1971), p. 147.

37. Letter of Tycho Brahe to Kepler, dated April 1, 1598, in GW XIII, nr. 92: ll. 90–96, tr. KY, p. 160.

38. Kepler's letter to Maestlin, written on October 3, 1595, in GW XIII, nr. 23: ll. 54–57.

39. Jasper Hopkins, *A Concise Introduction to the Philosophy of Nicholas of Cusa*, 2d ed. (Minneapolis: University of Minnesota Press, 1980), p. 113.

40. Ibid., p. 111.

41. See GW VI, pp. 218: ll. 29ff.

42. GW VI, p. 219: ll. 16–17.

43. Ibid., p. 219, Kepler's marginal note.

44. See GW XV, nr. 409: ll. 146ff.

45. Pauli, "Influence of Archetypal Ideas," p. 195.

46. Ibid., p. 194.

47. *Apologia*, GW VI, p. 446: ll. 20–21.

48. See "Kepler's Harmonies," BB, p. 528.

49. Burtt, *Metaphysical Foundations*, p. 64.

50. Idem.

51. These points we shall discuss in part 2.

52. Kepler's letter to Maestlin, written on August 2, 1595, in GW XIII, nr. 21: l. 9. See also MC, pp. 55, 157, and 221.

53. For a more detailed discussion of Kepler's work on musical harmony, especially in connection with planetary motions, see Owen Gingerich, "Kepler, Galilei, and the Harmony of the World," in his *Ptolemy, Copernicus, and Kepler* (New York: American Institute of Physics, 1991).

54. MC, p. 157.

55. GW III, p. 236: ll. 17–20, tr. KY, p. 188.

56. *Species* was a general idea found in the medieval tradition and in Renaissance thinkers like Kepler. Our astronomer did not explain exactly what the *species* was. According to him, in this context, it was something that came out of the material body of the sun. It was similar to both light and magnetism, but not identical with either one.

57. Even if one were to argue that Aristotle had this dynamic idea of order in some form, it is quite clear that he did not emphasize it. The fact that he considered any motion which deviated from "natural motion" as "violent motion" is quite revealing in this context.

58. See p. 75. However, Kepler later realized that this was not true and recanted his claim in the preface to his *Rudolphine tables*.

59. See Caspar, *Johannes Kepler*, p. 376.

60. Westman, "Kepler's Adoption of Copernicanism," p.184.

61. There is some controversy about whether or to what extent Copernicus subscribed to such a separation. In any case, there is no doubt that Kepler went far beyond what his master was willing to go.

62. Quoted by Caspar, *Johannes Kepler*, p. 373.

63. Ibid., p. 288.

64. Ibid., p. 267.

65. GW VI, p. 480.

66. GW XV, nr. 357: ll. 91–92.

67. See also Gingerich, "Kepler, Galilei, and the Harmony of the World," pp. 18-19.

68. GW VI, p. 329: ll. 13–15, tr. GB, p. 1049.

69. See Plato, *Timaeus* 36Bff. Also see *Republic* 617A and 617B.

70. See Aristotle, *De Caelo*, 290B, in *The Works of Aristotle*, tr. W. D. Ross (Oxford: Oxford University Press, 1947).

71. GW VI, p. 289: ll. 24–27.

72. Ibid., p. 289: ll. 30–31.

73. Westman, "Kepler's Theory of Hypothesis and the 'Realist Dilemma'," *Studies in History and Philosophy of Science* 3 (1972), 252.

74. Curtis Wilson, "Horrocks, Harmonies, and the Exactitude of Kepler's Third Law," *Science and History*, Studia Copernicana 16 (Ossolineum, 1978), p. 238.

75. Holton, "Johannes Kepler's Universe," p. 349.

76. See GW VI, pp. 211: l. 17–217: l. 23.

77. Kepler believed that the insensible or pure harmonies existed in the mind serving as models for the sensible harmonies.

78. GW VI, p. 323: ll. 9–13.

79. See Bruce Brackenridge, "Kepler, Elliptical Orbits, and Celestial Circularity: A Study in the Persistence of Metaphysical Commitment," part 2, *Annals of Science* 39 (1982), 285.

80. See idem.

81. GW VII, p. 275: ll. 8–12, tr. KY, p. 452.

82. This complementarity can be seen in yet another way also: the divine plan made use of geometry (theory of the regular solids) to fix the number of the planets and the principles of harmony to determine the eccentricities and periods of the planets.

83. GW VI, p. 361: ll. 39–41.

84. GW VI, p. 362: ll. 1–8, tr. KY, p. 342.

85. See p: 184: ll. 1–4.

86. For instance, GW III, p. 22: ll. 37–38.

87. GW III, p. 23: ll. 10–11.

88. MC, p. 155.

89. Kepler's letter to Erzherzog Ferdinand in July 1600, in GW XIV, nr. 166: ll. 56–57.

3
KEPLER'S SCIENTIFIC IDEAS

1. Kepler's letter to Brengger on October 4, 1607, GW XVI, nr. 448: ll. 4–6.

2. Ibid., nr. 448: ll. 8–10. I have translated the term *arithmeticam* as "mathematics." Although literally it should be "arithmetic," I think my translation is more faithful to what Kepler is conveying.

3. GW VII, p. 254: ll. 40–42, tr. GB, p. 850.

4. Newton said that Kepler got the laws by guesswork, rather than from observational study.

5. Max Caspar, *Johannes Kepler*, p. 130.

6. Plato, *Republic*, 529D and 530C, tr. Olaf Pederson, *Early Physics and Astronomy* (London: MacDonald and James, 1974), p. 27.

7. Plato, *Timaeus* 47A, tr. F. M. Cornford, *Plato's Cosmology* (London: Routledge and Kegan Paul, 1956), p. 157.

8. Kepler, GW VI, p. 226: ll. 21–22. Also in *Opera Omnia*, vol. 5, p. 224.

9. Kepler's letter to Fabricius on July 4, 1603, in GW XIV, nr. 262: ll. 129–130.

10. GW III, p. 191: l. 11.

11. Kepler's letter to Fabricius on October 1, 1602, in GW XIV, nr. 226: ll. 394–395.

12. For instance, see GW III, p. 124: l. 17–129: l. 35.

13. Kepler suffered from myopia and multiple vision.

14. Kepler's letter to Herwart on December 16, 1598, in GW XIII, nr. 107: ll. 135–136.

15. See Kepler's letter to Galileo on October 13, 1597, in GW XIII, nr. 76: ll. 55–70.

16. Kepler's letter to Herwart on March 26, 1598, in GW XIII, nr. 91: ll. 150–151.

17. Kepler's letter to Herwart on April 10, 1599, in GW XIII, nr. 117: ll. 19–22, tr. SM, p. 110.

18. GW XV, nr. 335: ll. 10–12.

19. Kepler's letter to Herwart on July 12, 1600, in GW XIV, nr. 168: ll. 102–104.

20. Kepler's letter to Maestlin on February 26, 1599, in GW XIII, nr. 113: ll. 89–92.

21. Tycho's letter to Kepler on December 9, 1599, in GW XIV, nr. 145: ll. 197–200.

22. GW XIII, nr. 94: ll. 18–22, tr. KY, p. 161.

23. In fact, he confessed to Herwart: "One of the most important reasons for m visit to Tycho was my desire to learn from him more correct figures for the eccentricities, by which I could check my *Mysterium* and the just mentioned *Harmony*. For these speculations a priori must not conflict with clear experimental evidence. Indeed they must be in conformity with it" (letter to Herwart on July 12, 1600, in GW XIV, nr. 168: ll. 105–109).

24. GW III, p. 272: ll. 1–2.

25. See GW VII, p. 318: ll. 20ff.

26. Kepler's letter to Brengger on November 30, 1607, in GW XVI, nr. 463: ll. 88–89.

27. Kepler's letter to Brengger on April 5, 1608, in GW XVI, nr. 488: ll. 348–349.

28. Examples of such terms are "force," "field," "atom," "gene," "subconscious," "drive," etc. They refer to certain entities which are introduced into scientific theories to explain phenomena under consideration. What is their ontological status? How are they related to observable factors? How are they verified? These are much talked-about issues in twentieth-century philosophy of science. My own concern is to discuss what place such entities had in Kepler's system.

29. GW XVI, nr. 438: ll. 18–19, tr. KY, p. 263.

30. GW III, p. 242: ll. 29–30, tr. *Great Ideas Today*, p. 326.

31. See ibid., p. 244: ll. 12–13.

32. See ibid., p. 243: ll. 30–31.

33. GW III, p. 241: ll. 10–13. See also Max Jammer, *Concepts of Force* (New York: Harper Torch Books, 1962), p. 87.

34. MC, p. 203.

35. Jammer, *Concepts of Force*, p. 90.

36. Arthur Koestler, "Kepler, Johannes," *The Encyclopedia of Philosophy*, vol. 4 (New York: Macmillan, 1972), p. 333.

37. GW III, p. 241: ll. 5–23, tr. Koestler, *Watershed*, p. 157.

38. Concerning the Neoplatonic roots of this idea of force, see Fritz Krafft, "The New Celestial Physics of Johannes Kepler," p. 195.

39. Kepler's argument goes as follows: Just as in the trinitarian sphere the Holy Spirit fills the space between the center and the surface, so also in the sphere of the universe the *species* or emanation fills the space between the fixed stars and the sun. Just as the Holy Spirit is ever active and dynamic, so too the emanation in the intervening space is active and dynamic. Just as the Holy Spirit can affect the universe and bring about changes, so too the emanation in the intervening space can exert an influence on bodies and cause changes in them.

40. GW III, p. 244: ll. 20–21, tr. *Great Ideas Today*, p. 327.

41. GW VII, p. 333: ll. 2–4. See also GW III, p. 244: ll. 20–21.

42. MC, p. 171.

43. GW III, p. 25: ll. 32–36. See Jammer, *Concepts of Mass* (Cambridge, Mass.: Harvard University Press, 1961), p. 54.

44. GW VII, p. 301: ll. 22–25. Also *Opera Omnia* vol. 6, p. 346.

45. Literally the last part of the quotation reads as follows: "with respect to the motive force, the inertia of matter is not like nothing to something" (GW VII, pp. 296: l. 42–297: l. 2). For the translation, see Jammer, *Concepts of Mass*, p. 55.

46. Owen Gingerich, "Kepler, Johannes," *Dictionary of Scientific*

Biography, vol. 7 (New York: Charles Scribner's, 1973), p. 292.

47. GW XVII, nr. 750: ll. 194–197.

48. See Kepler's letter to Herwart on February 10, 1605, in GW XV, nr. 325: ll. 60–61.

49. Gerald Holton, "Johannes Kepler's Universe," p. 346.

50. Ibid., p. 347.

51. Also it may be noted that because of the problem of "action at a distance," Newton and others did not regard the Newtonian forces as giving a fully satisfactory explanation.

52. Galileo's theory of the tide was incorrect. For one thing, unlike Kepler, he refused to attribute any definite role to the moon. Also, his theory could predict only one high tide daily. See Stillman Drake, *Galileo Studies* (Ann Arbor: University of Michigan Press, 1970), p. 205.

53. For Kepler's explanation of the tide, see GW III, p. 26: ll. 1–23.

54. See GW XIV, nr. 248: ll. 240ff.

55. See ibid., nr. 248: ll. 350–359.

56. Ibid., nr. 248: ll. 379–384.

57. This point we have discussed in chapter 1. See Kepler's comment in Fabricius's letter, dated February 7, 1603, in GW XIV, nr. 248: ll. 375–377.

58. Kepler's letter to Fabricius on July 4, 1603, in GW XIV, nr. 262: ll. 421–422. Here Kepler refers to the astrological beliefs of Fabricius, who was also a professional astrologer.

59. Letter written on February 28, 1624, in GW XVIII, nr. 974: ll. 211–213, tr. CB, p. 149.

60. Letter on September 21, 1616, in GW XVII, nr. 744: ll. 24–30.

61. Ibid., nr. 744: ll. 77–78.

62. Ibid., nr. 744: ll. 92–93.

63. GW XIII, nr. 43: ll. 6–10, tr. CB, p. 37.

64. MC, pp. 96–99.

65. Kepler's letter to Maestlin on October 3, 1595, in GW XIII, nr. 23: ll. 203–204, tr. CB, p. 31.

66. See CB, p. 30.

67. See Jardine, *Birth of History and Philosophy of Science*, p. 250.

68. Kepler's letter to Maestlin on October 3, 1595, in GW XIII, nr. 23: ll. 204–205, tr. CB, p. 31.

69. See GW XIV, nr. 168: ll. 108–109.

70. See Jürgen Mittelstrass, "Methodological Elements of Keplerian Astronomy," *Studies in History and Philosophy of Science* 3 (1972), 222.

71. See MC, p. 173.

72. It may be objected that Kepler's investigation of Mercury was a counterexample to my claim that he was using the H–D method, since if he really adhered to that method, he should have discarded the whole hypothesis because it failed to account for Mercury. However, this kind of problem is not peculiar to Kepler alone. In fact, this objection points to a general criticism of the H–D method, namely, in actual practice scientists do not discard a hypothesis simply because it fails to account for a particular result. They usually propose auxiliary hypotheses to modify the original one. Just because they employ auxiliary hypotheses, we do not say that they have given up the H–D method. It shows only that the H–D method has to be modified, and that this modified form of the method is more satisfactory. Kepler's study of Mercury in this context shows that he already had discovered this modified form of the H–D method.

73. It is true that not all philosophers of science believe that the hypothesis in the H–D method is obtained from irrational (or nonrational) or psychological sources. However, the rational basis of such a hypothesis is still a matter of controversy.

74. Kepler's letter to Fabricius on July 4, 1603, in GW XIV, nr. 262: ll. 127–129.

75. Ibid., nr. 262: ll. 158–159.

76. MC, p. 65.

77. This Keplerian use of chance will appear again in part 2, in connection with an important step in the discovery of the ellipse.

78. Idem.

79. Letter to Fabricius on February 7, 1603, in GW XIV, nr. 248: ll. 487–489.

80. Letter to Fabricius on July 4, 1603, in GW XIV, nr. 262: ll. 129–131.

81. These are mainly the principles discussed in chapters 1 and 2.

82. MC, p. 65.

83. Ibid., pp. 65–67.

84. The *Epitome*, although important, is not discussed here, because it is basically a systematic summary of all his major works.

85. In the case of this theory he seems to be less particular about observational testing.

86. See GW VI, pp. 323: l. 14–324: l. 15.

87. However in his letter to Matthias Bernegger on June 30, 1625, he reported that he had arrived at the moment of creation by calculation (see GW XVIII, nr. 1010: ll. 30–38).

4
THE ACCEPTANCE OF COPERNICANISM

1. Letter to Herwart on March 26, GW XIII, nr. 91: ll. 188–192.

2. GW VII, p. 8: ll. 26–29, tr. Burtt, *Metaphysical Foundations*, p. 58.

3. Letter of Bellarmine to P. A. Foscarini on April 12, 1615, in *Opere di Galileo*, vol. 12, p. 71, tr. in R. Blake, C. J. Ducasse, and E. Madden, *Theories of Scientific Thought* (Seattle: University of Washington Press, 1960), pp. 44–45 (emphasis added).

4. Osiander "To the Reader on the Hypotheses in This Work," in *Copernicus: On the Revolutions of the Heavenly Spheres*, tr. A. M. Duncan (New York: Barnes and Noble, 1976), p. 22.

5. ER, p. 7.

6. DR, book 1, ch. 8, tr. ER, p. 16.

7. Preface to DR, tr. ER, p. 4 (emphasis added).

8. Preface to DR.

9. Kuhn, *Copernican Revolution*, p. 137.

10. Preface to DR, tr. ER, p. 5.

11. Neugebauer, "On the Planetary Theory of Copernicus," in *Vistas in Astronomy* 10, ed. Arthur Beer (Oxford: Pergamon Press, 1968), p. 100.

12. Kuhn, *Copernican Revolution*, p. 134.

13. Idem.

14. See Hanson, *Conjectures*, p. 229.

15. See preface to DR, tr. ER, p. 5.

16. Norwood Hanson, "The Copernican Disturbance and the Keplerian Revolution," *Journal of the History of Ideas* 22 (1961), 184.

17. According to Plato, only uniform circular motion was permitted in the heavens. All observed noncircular motions would have to be accounted for in terms of uniform circular motion.

18. Ludwig Prowe, *Nicolaus Copernicus*, vol. 2, p. 176, quoted and tr. Koestler in *Sleepwalkers* (New York: Macmillan, 1959), p. 200.

19. See Caspar, *Johannes Kepler*, pp. 40–41.

20. See Kepler's letter to Maestlin on December 22, 1616, in GW XVII, nr. 750: ll. 246–247.

21. Westman, "Kepler's Adoption of Copernicanism," p. 23.

22. Idem.

23. Ibid., p. 74.

24. Ibid., p. 55.

25. See ibid., p. 69.

26. Idem. See also Robert Westman, "The Comet and the Cosmos: Kepler, Maestlin and the Copernican Hypothesis," in *The Reception of Copernicus' Heliocentric Theory*, ed. Jerzy Dobrzycki (Dordrecht: Reidel, 1972), p. 27.

27. Westman, "Kepler's Adoption of Copernicanism," p. 68.

28. See Derek Price, "Contra Copernicus," in *Critical Problems in the History of Science*, ed. Marshall Clagett (Madison: University of Wisconsin Press, 1969), pp. 197–218.

29. See Kristian Moesgaard, "Copernican Influence on Tycho," in *The Reception of Copernicus' Heliocentric Theory*, ed. Dobrzycki, p. 37.

30. For instance, Kepler wrote: "We shall demonstrate both here and throughout this book the most perfect, geometrical equivalence of the three forms [those of Ptolemy, Tycho Brahe, and Copernicus]" (AN, p. 89: ll. 12–13; see also p. 87: ll. 15–19).

31. See MC, p. 87.

32. Westman, "Kepler's Adoption of Copernicanism," p. 73.

33. MC, p. 79.

34. GW XIII, nr. 23: ll. 45–52.

35. Westman, "Kepler's Adoption of Copernicanism," p. 74.

36. Although Burtt's book was first published in 1924, it generated and still generates considerable interest. This is evident from the fact that it was reprinted several times. The Humanities Press edition came out in 1952 and the Anchor Books revised edition in 1954. Burtt's arguments and ideas about the role that religious views played in Kepler's acceptance of Copernicanism are among the most notable of their kind.

37. Burtt, *Metaphysical Foundations*, p. 58.

38. Ibid., p. 60.

39. Kepler, *Opera Omnia*, vol. 8, p. 267, tr. Burtt, *Metaphysical Foundations*, p. 59.

40. See MC, pp. 75–85.

41. See GW XIII, nr. 23: ll. 45–52.

42. See GW XV, nr. 340: ll. 288–381.

43. Gingerich believes that, in addition to empirical reasons, religious reasons also played a role in Kepler's acceptance of the new theory. Accordingly, "Kepler's personal introduction to the Copernican system bordered on a religious experience" ("Kepler as a Copernican," in *Johannes Kepler—Werk und Leistung* [Linz, 1971], p. 111.) However, he does not develop this idea of religious experience to show exactly how it contributed towards Kepler's acceptance of Copernicanism.

44. MC, p. 75.

45. See Aiton's note in MC, p. 235. Also see Gingerich, " 'Crisis' versus Aesthetic in the Copernican Revolution," *Vistas in Astronomy*

17, ed. A. Beer and K. Strand (Oxford: Pergamon Press, 1975), pp. 85–93. For a translation by Gingerich and W. Walderman of the preface to the *Rudolphine tables*, see *Quarterly Journal of the Royal Astronomical Society* 13 (1972), 360–73.

46. See also A. C. Crombie, *Medieval and Early Modern Science* 2 (Cambridge, Mass.: Harvard University Press, 1967), p. 180.

47. Although it cannot be considered as supplying a necessary and sufficient condition for the adoption of the theory, it must certainly be considered as one of the reasons.

48. MC, p. 79.

49. Quoted by Westman, "Kepler's Adoption of Copernicanism," p. 55.

50. Quoted by Westman, "The Comet and the Cosmos," p. 23.

51. GW XIII, nr. 23: ll. 49–51.

52. MC, p. 75.

53. See MC, p. 81.

54. Westman, "Kepler's Theory of Hypothesis and the 'Realist Dilemma,'" *Studies in History and Philosophy of Science* 3 (1972), 237.

55. See Kuhn, *Copernican Revolution*, pp. 168–69.

56. Gingerich, "Ptolemy, Copernicus, and Kepler," in *Great Ideas Today*, p. 140.

57. See Gingerich, "'Crisis versus Aesthetic," p. 90.

58. Hanson, *Conjectures*, p. 229.

59. See A. Pannekock, *History of Astronomy* (New York: Interscience Pub., 1961), p. 223. See also Moesgaard, "Copernican Influence on Tycho," pp. 48–53.

60. See Pannekock, *History of Astronomy*, p. 223. See also BB, p. 202.

61. MC, p. 77.

62. Idem.

63. Idem.

64. MC, pp. 79–81.

65. DR, book 1, ch. 10, tr. Hanson, *Conjectures*, pp. 188–89.

66. MC, p. 77.

67. MC, pp. 75–77.

68. Preface to DR, tr. Hanson, *Conjectures*, p. 182.

69. MC, p. 77.

70. MC, p. 75.

71. One may argue that Kepler probably did not know of these counterclaims when he accepted the new theory. This seems to be quite improbable since at that time the opposition to Copernicus was very severe and all these objections were very much talked about.

72. See Kuhn, *Copernican Revolution*, p. 170. Note that figure

3 is not accurately shown to scale. The deviation of the sun from the center is exaggerated. As Dreyer points out, if the radius of the earth (O_E) is 1, then SO is 0.0368 and OO_E is 0.0047 (see Dreyer, *History of Astronomy*, p. 332).

73. DR, book 1, ch. 10, tr. Hanson, *Conjectures*, p. 189.

74. Kepler used a similar argument to assert that the interplanetary space was not made up of any solid material (see MC, p. 155).

75. MC, p. 63.

76. Letter written on October 3, 1595, in GW XIII, nr. 23: ll. 83–84.

77. Letter to Herwart on March 28, 1605, in GW XV, nr. 340: ll. 320–324.

78. Only the religious could argue for the unique position of the sun in the theory; only the philosophical could ensure that the theory satisfied first principles; only the empirical could sufficiently establish the theory's observational validity.

5
THE DEVELOPMENT OF A TRULY HELIOCENTRIC VIEW

1. ER, p. 20.

2. DR, book 1, ch. 10, tr. ER, p. 22.

3. For instance see E. J. Dijksterhuis, *The Mechanization of the World Picture*, tr. C. Dikshoorn (Oxford: Clarendon Press, 1961).

4. Letter written on March 5, 1605, in GW XV, nr. 335: ll. 62–64. The same idea can be found in GW III, p. 79: ll. 9–12.

5. AN, p. 65: ll. 32–36, tr. William Donahue, *New Astronomy* (Cambridge: Cambridge University Press, 1992).

6. Gingerich, "Kepler's Place in Astronomy," BB, p. 264.

7. RS, p. 154.

8. See GW III, p. 65: ll. 33–34.

9. Kepler's letter to Herwart on July 12, 1600, in GW XIV, nr. 168: ll. 120–124.

10. Kepler's letter to Fabricius, in February, 1604, GW XV, nr. 281: ll. 158–162.

11. GW III, p. 337: ll. 14–18.

12. GW III, p. 91: ll. 18–20.

13. Kepler's letter to Fabricius on December 2, 1602, in GW XIV, nr. 239: ll. 154–156.

14. See GW III, 238: ll. 12–29.

15. GW II, p. 19: ll. 37–41, tr. Pauli, "Influence of Archetypal Ideas," p. 170.

16. GW III, p. 20: ll. 24–26, tr. Donahue, *New Astronomy*.

17. GW III, p. 96: l. 15.

18. This proof also relied on the results of chapter 51, where he established that when the planet occupied corresponding positions on the opposite semicircles (or semiovals) it was equidistant from the sun.

19. AN, p. 96. Here also *a priori* is understood as being opposed to *empirical*.

20. See GW III, pp. 133: l. 19–141: l. 40.

21. RS, p. 169.

22. See GW XIV, nr. 190: ll. 120–125.

23. GW III, p. 141: ll. 14–15.

24. Ibid., p. 141: ll. 15–16.

25. "Impertinenti diversorum orbium colligatione" (GW III, p. 141: l. 13).

26. "Quod monstri simile sit" (GW III, p. 141: l. 11).

27. It may be noted that although in the geocentric system the ecliptic is the plane of the sun's orbit, in the heliocentric system, it is the plane of the earth's orbit. Yet the ecliptic has been closely associated with the sun since it is usually understood as the apparent annual path of the sun on the celestial sphere. As we can see from Kepler's writings, there is no doubt that this association was very active in his mind at this stage of his work.

28. See MC, pp. 217–219.

29. MC, p. 216.

30. MC, p. 217.

31. MC, p. 219.

32. See letter to Herwart on July 12, 1600, in GW XIV, nr. 168: ll. 127–128.

33. GW III, p. 97: ll. 39–40.

34. Moesgaard, "Copernican Influence on Tycho," p. 38.

35. See MC, p. 219.

6
THE VICARIOUS HYPOTHESIS
AND ITS FAILURE

1. See RS, p. 180.

2. Since Kepler refused to accept Ptolemy's bisection of the eccentricity and wanted to determine the position of the equant solely on the basis of observations, he needed at least four observations at opposition.

3. It may be noted that Tycho Brahe through his ingenious method could get an accuracy up to 2 minutes. Hence the traditional limit of accuracy was modified since the time of Tycho.

4. An error of 8 minutes hardly disturbed any one of them, except Tycho; that was good astronomy, as far as they were concerned.

5. GW III, p. 178: ll. 10–12, tr. KY, p. 179.

6. GW XV, nr. 323: ll. 362–366.

7. See GW III, 182: ll. 25–28.

8. GW III, p. 182: ll. 27–28.

9. For instance, the principle of geometrizability had its root in God the geometer; the principle of precision, in God the rational; the principle of simplicity, in God the simple (the absolutely one and undivided); and the principle of economy, in God, the *optimus creator*, who allowed absolutely no waste in the created universe.

10. GW III, p. 178: ll. 1–6, tr. KY, p. 178. I have modified the translation.

7
THE FINAL BREAK WITH GEOCENTRISM

1. MC, p. 219.

2. MC, p. 105.

3. See idem.

4. MC, p. 238.

5. Letter written on April 10, 1599, in GW XIII, nr. 117: ll. 158–159, tr. SM, p 136.

6. See AN, p. 191: ll. 19–26.

7. Letter dated July 12, 1600, in GW XIV, nr. 168: ll. 127–128.

8. Kepler's letter on June 1, 1601, in GW XIV, nr. 190: ll. 59–62.

9. See RS, p. 199.

10. For details see RS, pp. 199–201.

11. Tycho's letter to Kepler on April 1, 1598, in GW XIII, nr. 92: ll. 51–54. Also in GW III, p. 191: ll. 29–32, tr. Donahue, *New Astronomy*. See RS, p. 199.

8
THE DISCOVERY OF THE SECOND LAW

1. KY, p. 121.

2. See GW III, p. 69: ll. 1–18.

3. KY, p. 121.

4. KY, p. 419.

5. See idem.

6. In fact, Ptolemy did not postulate any relation between distance and speed, but his equant implied that in the line of apsides an inverse relation existed between the two.

7. KY, p. 185. I repeat that the statement is expressed in modern rather than Keplerian terms. For instance, Kepler lacked our modern idea of instantaneous velocity. Also, as I indicate in the next paragraph, he often talked of the delay (*mora*), or time required or spent by a planet while traversing a given arc of its path, rather than of its speed or velocity.

8. For the actual working out of this proof see RS, pp. 210–214.

9. See KY, pp. 318–319 and 404.

10. In this case, the planet could be sometimes farther, sometimes closer, than the center of the epicycle. The actual distance of the planet did not matter; what mattered was the distance of the empty center of the epicycle.

11. Since the idea of the distance law was already around, strictly speaking it may not be quite accurate to talk of this as Kepler's discovery.

12. See MC, p. 199.

13. See GW III, p. 233: ll. 22–28.

14. GW III, p. 236: ll. 8–13.

15. Ibid., p. 236: ll. 14–16.

16. MC, p. 199.

17. See GW III, p. 236: ll. 17–20.

18. Ibid., p. 236: ll. 27–29.

19. Ibid., p. 236: ll. 32–33.

20. See GW III, p. 69: ll. 17–18.

21. GW III, p. 237: ll. 7–10.

22. Ibid., p. 237: ll. 11–13.

23. GW III, p. 238: ll. 12–14.

24. See MC, p. 199.

25. The a priori considerations he developed both in HM and in the *Epitome*.

26. GW III, p. 238: ll. 15–20. In part 1, I have pointed out that Kepler's a priori method possessed practically all the main characteristics of the H–D method. Here he points out that the a priori method could start from his beliefs on the nature and nobility of the sun. Since such beliefs could be an integral part of his hypothesis, there need not be any inconsistency. The a posteriori method he talks of seems to be one without any hypothesis whatever, being totally dependent on observations only.

27. AN, p. 24: ll. 9–10.

28. Kepler's letter to Herwart on March 28, 1605, in GW XV, nr. 340: ll. 340–342, tr. CB, p. 74.

29. Letter to Herwart on December 16, 1598, in GW XIII, nr. 107: ll. 157–159.

30. GW III, p. 242: ll. 13–14.

31. MC, p. 199.

32. Ibid., p. 201.

33. It may be noted that Kepler first developed this method in the course of his study of the orbit of the earth and later applied it to that of Mars. Thus he wrote about the motion of the earth: "After that, for the period of revolution, although the exact figure should be 365 days and 6 hours, I assigned another, round value, and supposed it to correspond to 360 degrees, or the whole circle, which the astronomers call the mean anomaly" (GW III, p. 263: ll. 30–33).

34. GW III, p. 264: ll. 1–3.

35. GW III, p. 264: ll. 3–7. See KY, p. 233.

36. Detailed discussion of this procedure and discovery can be found in Small, Koyré, etc. Since I focus here mainly on the factors responsible for this discovery, I have chosen to remain brief about these points.

37. KY, p. 232.

38. Letter dated July 4, 1603, in GW XIV, nr. 262: ll. 41–43.

39. GW III, p. 267: ll. 3–4.

40. Ibid., p. 268: ll. 9–10.

41. See GW III, p. 268: ll. 8–9.

42. Stephenson, *Kepler's Physical Astronomy*, p. 83.

43. Letter written on July 4, 1603, in GW XIV, nr. 262: ll. 38–40.

9
THE DISCOVERY OF THE FIRST LAW

1. Hanson, *Conjectures*, p. 230.

2. Brackenridge, "Kepler, Elliptical Orbits, and Celestial Circularity," p. 120.

3. See Hanson, "Copernican Disturbance," p. 174.

4. With the decline of the Greek influence, this knowledge was very much lost to most people. For instance, many Church Fathers did not believe in a spherical earth. In DR Copernicus refers to Lactantius, one of the outstanding Church Fathers, as one of his opponents (see ER,

p. 5). However, the professional astronomers and the Aristotelians must have been well acquainted with this idea of a spherical earth.

5. Aristotle, *De Caelo*, 269A, tr. W. D. Ross.

6. Hanson, "Copernican Disturbance," p. 174.

7. Aristotle, *De Caelo* 286A, tr. W. D. Ross.

8. Tr. Kuhn, *Copernican Revolution*, p. 147.

9. Idem.

10. Letter written on December 9, 1599, in GW XIV, nr. 145: ll. 206–208.

11. Ibid., nr. 145: l. 205.

12. Ibid., nr. 145: ll. 202–205, tr. KY, p. 162.

13. MC, p. 97.

14. AN, p. 61: ll. 5–8.

15. See KY, pp. 416 and 419.

16. See GW III, p. 282: ll. 6–10.

17. See GW III, p. 283: ll. 13–22.

18. For a brief discussion of this case see RS, pp. 238 and 110–111.

19. GW III, p. 284: ll. 7–10.

20. See GW III, pp. 285: ll. 14–287: ll. 36.

21. See ibid., p. 286: ll. 10–12.

22. GW III, p. 286: ll. 13–16.

23. GW III, pp. 286: ll. 33–35 and 287: ll. 35–36. See KY, p. 244.

24. For instance, the textbook writer Eric Rogers believed that Kepler found the shape of the earth's orbit by plotting the different values. See Curtis Wilson, "Kepler's Derivation of the Elliptical Path," *Isis* 59 (1968), 5. Here the orbit of the earth was under consideration. But the same could be said in the case of Mars as well.

25. See GW XV, nr. 297: ll. 35–80.

26. Kepler's letter to Fabricius on December 18, 1604, in GW XV, nr. 308: ll. 67–71.

27. Ibid., nr. 308: ll. 71–72.

28. GW III, p. 255: ll. 2–5.

29. Ibid., p. 255: ll. 6–10.

30. Ibid., p. 255: ll. 35–39.

31. GW III, p. 257: ll. 9–11.

32. Ibid., p. 257: ll. 26–29.

33. GW III, p. 258: l. 19.

34. KY, p. 222.

35. See KY, p. 220.

36. GW III, p. 258: ll. 8–10.

37. In fact, initially Kepler was very much in favor of such minds

or intelligences since he himself wrote in AN: "I deny that any perennial, non-rectilinear motion that is not governed by a mind has been created by God" (p. 69: ll. 17–18).

38. See GW III, p. 262: ll. 23–24. See also KY, pp. 223 and 406. Animal souls or motive souls discussed here could give rotatory motion, but not translatory motion.

39. GW XV, nr. 323: ll. 291–294.

40. Kepler wrote in AN: "Whichever method discussed above is employed to delineate the path of the planet, it follows that the path is truly an oval, not an ellipse" (GW III, p. 295: ll. 22–24).

41. It is sometimes suggested that Kepler's idea of the oval was inspired by the ancient mythical cosmogony of the universe coming from a cosmic egg. I have not found any evidence in Kepler's writings to substantiate such a suggestion. It is indeed true that many ancient cosmogonies do talk about the universe originating from an egg, e.g., the Indian mythology. Again, since Aristotle argued that a spherical body should execute circular motion, it is plausible to argue that an egg-shaped object should move in an oval path. However, it may be noted that the mythology does not say that the universe is oval in shape, rather it says that the universe came out of an egg. Hence, logically speaking, the Aristotelian argument is not applicable here. In any case, there is no evidence that Kepler ever entertained this idea or this kind of argument.

42. H. W. Turnbull, ed., *The Correspondence of Isaac Newton* 2 (Cambridge: Cambridge University Press, 1960), p. 436.

43. See above, From the Circle to the Oval, Factors Responsible for This Discovery, Philosphical Ideas. See also KY 247.

44. GW III, p. 289: ll. 22–26.

45. See GW III, p. 289: ll. 37–46. I have translated *aequidistans* as "parallel" since such a translation seems to bring out the idea better.

46. See GW III., p. 294.

47. GW III, p. 295: ll. 6–8.

48. Ibid., p. 295: ll. 22–24.

49. See GW III, p. 299: ll. 11–17.

50. See GW III, p. 302: ll. 30–40.

51. See ibid., p. 303: l. 10.

52. GW III, p. 309: ll. 36–39.

53. Ibid., p. 310: ll. 10–11.

54. RS, p. 257.

55. GW III, p. 310: ll. 23–31.

56. See RS, pp. 255–256 and 259.

57. GW III, p. 311: ll. 10–12.

58. GW, p. 313: ll. 29–34. There is a misprint in the table. Kepler

mistakenly gives 114 instead of 144.

59. Ibid., p. 313: l. 38; 314: l. 1.

60. Ibid., p. 314: ll. 8–10.

61. See RS, pp. 269–273. Donahue argues that the empirical basis for the table of distances given in chapter 53 of AN "is a fraud, a complete fabrication. . . . The observational input is nil" ("Kepler's Fabricated Figures," pp. 233–234 [see above, Introduction, note 1]). In one of my unpublished papers I have argued that in deciding this case we need to take into account many other relevant facts as well. It seems to me that when this is done, we will be led to a conclusion different from Donahue's.

62. See GW III, p. 344: ll. 5–13.

63. See GW III, chapter 51. See also RS, p. 265.

64. See GW III, p. 345: ll. 30–34.

65. Ibid., p. 345: l. 36.

66. Optical equation and physical equation were technical terms used by Kepler in his work. Simply put, we may say that the former arose as a result of the variation of distance between the planet and its eccentric center, and consequent variation in the apparent angular motion of the planet. The latter arose from the actual variation of the speed of the planet's real motion. The sum of the optical and physical equations gave the eccentric equation.

67. GW III, p. 346: ll. 2–7.

68. Ibid., p. 346: ll. 8–11, tr. KY, p. 253.

69. GW III, p. 347: ll. 7–9.

70. See RS, p. 279.

71. GW XV, nr. 335: ll. 67–69.

72. GW III, p. 346: ll. 14–18.

73. See GW III, p. 346: ll. 22–26.

74. Kepler's letter to Maestlin on March 5, 1605, in GW XV, nr. 335: ll. 90–91. The two false hypotheses Kepler referred to are the one about the circular orbit and the other about epicyclic motion.

75. Ibid., nr. 335: ll. 91–92.

76. GW III, p. 289: ll. 33–37.

77. Ibid., p. 349: ll. 1–6.

78. See ibid., p. 348: ll. 11ff.

79. Kepler's letter to Maestlin on March 5, 1605, in GW XV, nr. 335: ll. 93–94.

80. GW III, p. 350: ll. 4–5.

81. Kepler's letter to Maestlin on March 5, 1605, in GW XV, nr. 335: ll. 95–96, tr. KY, p. 252.

82. See AN, p. 289: ll. 35–36.

83. GW XV, nr. 335: ll. 99–101.

84. Kepler's letter to Maestlin on March 5, 1605, in GW XV, nr. 335: ll. 108–115, tr. KY, pp. 252–253.

85. AN, p. 352: ll. 19–21.

86. See GW III, pp. 355: ll. 31–41 and 362: ll. 1–9 for details.

87. Ibid., p. 355: ll. 20–22.

88. Ibid., p. 366: ll. 1–4.

89. GW XV, nr. 358: ll. 390–392.

90. See Aiton, "Kepler's Second Law of Planetary Motion," *Isis* 60 (Spring, 1969), 83.

91. GW III, p. 367.

92. Small does talk about Kepler's magnetic theory (RS, pp. 214–216), but the discussion is very incomplete and brief. He fails to bring out the relation between the boat analogy and magnet analogy.

93. See KY, pp. 263–264.

94. Concerning this point there is some controversy. See Donahue's "Kepler's Fabricated Figures."

95. See GW III, p. 311: ll. 10–12.

96. GW XV, nr. 358: ll. 317–324.

97. Incidentally, from what has been said it follows that the "accidental" discovery was not all that accidental, after all. It was not something that came to him from nowhere, by mere chance. He was thinking about the issue and was looking for a solution. The method of diametral distances he seemed to have obtained as a result of serious, well–thought-out reasoning. As we have seen, he called it a chance discovery probably because this particular result was not something he had anticipated.

98. For example, the flowing river is material, whereas the emanation was immaterial.

99. See GW III, p. 290: ll. 1–17.

100. Letter of March 5, 1605, in GW XV, nr. 335: ll 121–122.

101. Letter of March 28, 1605, in GW XV, nr. 340: ll. 320–324.

102. MC, p. 39.

103. GW III, p. 310: l. 32.

104. Ibid., p. 362: ll. 15–16.

105. Of course, this was not his only reason for moving away from explanation of natural phenomena in terms of spirits and souls. For other reasons, see above, part 1, chapter 3.

106. See Kepler's letter to Maestlin in GW XV, nr. 335: l. 90.

107. MC, p. 175.

108. Jardine, *Birth of History and Philosophy of Science*, pp. 140–141.

CONCLUSION

1. For instance, as I have argued, in the identification of the sun as the source of motion and cause of planetary motion, philosophical and religious principles played a dominant role; in the last stages of the discovery of the first law, empirical considerations were more conspicuous.

2. My point is not that he did not make use of this kind of method, but rather that his discovery process cannot be reduced to a straightforward H–D method.

3. Hanson, *Patterns of Discovery*, p. 86.

4. Ibid., p. 84.

5. See GW III, p. 243: l. 1.

6. MC, p. 85.

7. Kuhn, *Copernican Revolution*, p. 191.

8. James Brodrick, S.J., *The Life and Work of Blessed Robert Francis Cardinal Bellarmine, S.J.* (New York: P. J. Kennedy and Sons, 1928), p. 359.

9. Idem.

10. GW XIII, nr. 89: ll. 45–51.

11. Letter on March 28, 1605, in GW XV, nr. 340: ll. 85–86. See also GW II, p. 281: ll. 30–38.

12. See ibid., nr. 340: ll. 83–84.

13. See James Langford, *Galileo, Science, and the Church* (Ann Arbor: University of Michigan Press, 1971), p. 65.

14. GW VII, p. 99: ll. 27–29, tr. Rosen, "Kepler and the Lutheran Attitude," BB, p. 334.

15. GW III, p. 30: ll. 13–16, tr. *Great Ideas Today*, p. 319.

16. GW III, p. 30: ll. 1–4.

17. Here "demonstratively" means "proved beyond all doubt."

18. See Kuhn, *Copernican Revolution*, p. 191.

19. MC, p. 75.

20. Hanson, "Copernican Disturbance," p. 169.

21. Written on February 10, 1605, in GW XV, nr. 325: 11. 57–61, tr. Koestler, *Watershed*, p. 155.

22. See KY, p. 378.

23. Mittelstrass, "Methodological Elements of Keplerian Astronomy," p. 207.

24. Quoted in Jardin, *Birth of History and Philosophy of Science*, pp. 144–145.

25. Ibid., p. 145.

26. See Edward Grant, "Late Medieval Thought, Copernicus, and the Scientific Revolution."

27. Commenting on the changed perspective of mathematics in the seventeenth century Funkenstein writes: "What to the ancients was a cardinal vice, and to the Middle Ages a lesser one, now became a virtue" (*Theology and the Scientific Imagination*, p. 316). Although he overlooks Kepler's outstanding contribution towards this transition, our study emphasizes Kepler's role in this development.

28. GW III, p. 20: 11. 21–22, tr. *Great Ideas Today*, p. 312.

29. Written in the beginning of 1605, in GW XV, nr. 323: 11. 175–178.

30. Ibid., nr. 323: 11. 23–26.

31. See Kepler's letter to Maestlin, written on December 22, 1616, in GW XVII, nr. 750: 11. 168–190.

FIGURE 13

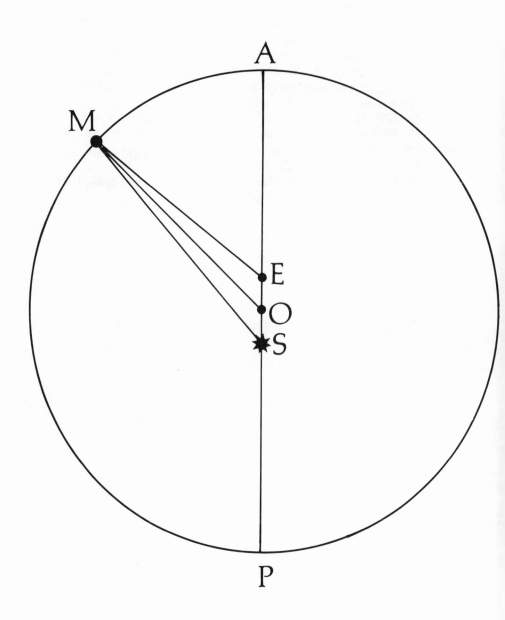

GLOSSARY OF SELECT TECHNICAL TERMS

In figure 13 (see GW XV, p. 173), S = the true sun, O = center of the planet's eccentric circle, M = position of the planet, E = equant.

ANOMALY. An angular value used to describe orbital motion of a celestial body. In this book we refer to three different kinds of anomaly.

1. TRUE or EQUATED ANOMALY: Angle ASM. The angle at the sun between the aphelion, the sun, and the planet.
2. MEAN ANOMALY. Angle AEM. The fraction of the period that has elapsed since the planet passed through aphelion, measured in angle.
3. ECCENTRIC ANOMALY. Angle AOM. The angle measured at the center of the eccentric circle, between the aphelion and the planet.

APHELION (A). The point on the orbit where a planet is at its greatest distance from the sun.

APSIS. Either of two points on an eccentric orbit of a planet, one farthest from the sun and the other nearest. In the figure, A and P are the apsides. The line AP is called the line of apsides.

CONJUNCTION. Two heavenly bodies are said to be in conjunction when they have the same longitude.

297

ECCENTRIC CIRCLE. The circular path that the sun was supposed to describe around the earth. Since the earth was placed at an off-center point, the path was called the eccentric circle. Copernicus, of course, displaced the earth with the sun.

ECLIPTIC. The apparent annual path of the sun on the celestial sphere.

EPICYCLE. Astronomers before Kepler believed that the true motion of a planet or of any other heavenly body had to be circular. However, this belief gave rise to a problem because observations showed marked deviation from circularity. The astronomers devised a method to resolve this problem. According to this device, an auxiliary circle or epicycle was placed on the original circle (deferent). The planet moved on the circumference of this epicycle, while the center of the epicycle moved on the circumference of the deferent. This device enabled the astronomers to account for observed deviations from circularity in terms of circular motion.

EQUANT POINT (E). The center of the circle around which the planet is supposed to move with uniform velocity. Often it is referred to simply as "equant."

INCLINATION. The angle between the plane of the planetary orbit and that of the ecliptic.

INEQUALITY. Any motion which deviated from a perfectly circular motion was called an inequality. In this book we talk of the first and the second inequalities.

1. THE FIRST INEQUALITY. From ancient times astronomers found that the planets and other bodies did not move with uniform circular motion around the central body. This deviation from uniform circular motion was termed the first inequality.
2. THE SECOND INEQUALITY. Astronomers noticed that planets executed retrograde motion, i.e., they first moved in one direction, their motion gradually slowed down until they came to a temporary stop; then they moved in the opposite direction with gradually decreasing velocity until they stopped motion temporarily once again. Afterwards they continued their motion around the central body. Thus the planetary motion was marked by loops. This deviation from uniform circular motion was called the second inequality.

LATITUDE. The planet's angular distance from the ecliptic. Depending on the point of observation we can have either heliocentric or geocentric latitude.

LONGITUDE. The angular distance between the planet's projection on the ecliptic and the vernal equinox. If the measurement is taken as viewed from the sun, we get heliocentric longitude, whereas if as viewed from the earth, we get geocentric longitude.

MAGNITUDE. A measure of a star's or planet's brightness. The higher the numerical value of the magnitude, the dimmer the star.

NODES. The points of intersection between the orbit of a celestial body and the ecliptic.

OPPOSITION. A planet is said to be in opposition when its position is such that it is in a plane containing the sun and the earth, at celestial longitude 180 degrees from the sun.

PARALLAX (STELLAR). The apparent change in the position of a star when observed from two different positions of the earth's orbit.

PERIHELION (P). Point on the orbit where the planet is nearest to the sun.

SYZYGY. A point in the orbit of a body at which it is in conjunction with or opposition to the sun.

ZODIAC. The portion of the spherical surface of the universe within which the apparent motions of the sun, moon, and planets take place. This surface is divided into 12 equal parts of 30 degrees each.

BIBLIOGRAPHY

Aiton, Eric J. "Kepler's Second Law of Planetary Motion." *Isis* 60 (1969), 75–90.

———. "Infinitesimals and the Area Law." In *Internationales Kepler-Symposium, Weil der Stadt*, 1971, ed. Fritz Krafft, Karl Meyer and Bernhard Sticker, pp. 285–305. Hildesheim, 1973.

———. "The Elliptical Orbit and the Area Law." In *Kepler: Four Hundred Years. Vistas in Astronomy* 18, ed. Arthur Beer and Peter Beer, pp. 573–583. Oxford: Pergamon Press, 1975.

———. "Johannes Kepler and the 'Mysterium Cosmographicum.' " *Sudhoffs Archiv* 61 (1977), 173–193.

Aristotle. *The Works of Aristotle*. Tr. W. D. Ross. Oxford: Clarendon Press, 1947.

Armitage, Angus. *John Kepler*. London: Faber, 1966.

Aquinas, Thomas. "Reason and Faith." *Summa Contra Gentiles*. Tr. The English Dominican Fathers. Reprinted in *Problems and Perspectives in Philosophy of Religion*, ed. G. I. Marodes and S. C. Hackett, pp. 10–19. Boston: Allyn and Bacon, 1967.

Bacon, Francis. *Novum Organum*. Ed. Thomas Fowler. Oxford: Clarendon Press, 1889.

Banville, John. *Kepler*. Boston: David R. Godine, 1983.

Baumgardt, Carola, ed. and tr. *Johannes Kepler: Life and Letters*. New York: Philosophical Library, 1951.

Beaulieu, Clement, S.S.C.C. "Kepler's Concept of Causality." Master's Dissertation, The Catholic University of America, Washington D.C., 1964.

Bechler, Zev, ed. *Contemporary Newtonian Research*. Dordrecht: Reidel, 1982.

Beer, Arthur. "Kepler's Astrology and Mysticism." In *Kepler: Four Hundred Years. Vistas in Astronomy* 18, ed. Arthur Beer and Peter Beer, pp. 399–426. Oxford: Pergamon Press, 1975.

Beer, Arthur, and Peter Beer eds. *Kepler: Four Hundred Years. Vistas in Astronomy* 18. Oxford: Pergamon Press, 1975. [This volume is referred to as BB.]

Bellarmine, Robert. *The Louvain Lectures of Bellarmine and the Autograph Copy of His 1616 Declaration to Galileo*. Tr. Ugo Baldini and George Coyne. Vatican Observatory Publications, 1984.

Bethune, Drinkwater J. *Life of Kepler*. In *Library of Useful Knowledge*. London: Baldwin and Cradock, 1830.

Bialas, Volker. "Keplers komplizierter Weg zur Wahrheit: Von neuen Schwierigkeiten, die 'Astronomia Nova' zu lesen." *Berichte zur Wissenschaftsgeschichte* 13 (1990), 167–176.

Black, Max. *Models and Metaphor*. Ithaca, N.Y.: Cornell University Press, 1962.

Blackwell, Richard J. *Galileo, Bellarmine, and the Bible*. Notre Dame, Ind.: University of Notre Dame Press, 1991.

Blake, R., C. J. Ducasse and E. Madden. *Theories of Scientific Method*. Seattle: University of Washington Press, 1960.

Brackenridge, Bruce. "Kepler, Elliptical Orbits, and Celestial Circularity: A study in the Persistence of Metaphysical Commitment," parts 1 and 2. *Annals of Science* 39 (1982), 117–143 and 265–295.

Brewster, David. *The Martyrs of Science*. New York: Harper, 1981.

Briggs, Robin. *The Scientific Revolution of the Seventeeth Century*. London: Longman, 1969.

Broad, William. "After 400 Years, a Challenge to Kepler: He Fabricated His Data, Scholar Says." *New York Times* (January 23, 1990), C1.

Buchdahl, Gerd. "Methodological Aspects of Kepler's Theory of Refraction." *Studies in History and Philosophy of Science* 3 (1972), 265–298.

Burke-Gaffney, M. W., S.J. "Johann Kepler and Modern Astronomy." Ph.D. Dissertation, Georgetown University, 1935.

———. *Kepler and the Jesuits*. Milwaukee: Bruce Publishing Company, 1944.

Burtt, E. A. *The Metaphysical Foundations of Modern Science*. New York: Doubleday Anchor Books, 1954.

Butterfield, H. *The Origins of Modern Science: 1300–1800*. New York: Macmillan, 1958.

Caspar, Max. *Johannes Kepler: 1571–1630*. Tr. Doris Hellman. New York: Abelard-Schuman, 1959.

Clagett, Marshall. *Studies in Medieval Physics and Mathematics*. London: Variorum Reprints, 1979.

Copernicus, Nicholas. *De Revolutionibus*. Reprint of the first edition of 1543. New York: Johnson Reprint, 1965.

———. *On the Revolutions*. Ed. Jerzy Dobrzycki. Tr. Edward Rosen. Baltimore: Johns Hopkins University Press, 1978.

Crombie, A. C. *Medieval and Early Modern Science* 2. Cambridge, Mass.: Harvard University Press, 1967.

DeKosky, Robert K. *Knowledge and Cosmos: Development and Decline of the Medieval Perspective*. Washington, D.C.: University Press of America, 1979.

Dijksterhuis, E. J. *The Mechanization of the World Picture*. Tr. C. Dikshoorn. Oxford: Clarendon Press, 1964.

Donahue, William. "Kepler's Fabricated Figures: Covering up the Mess in the *New Astronomy*." *Journal for the History of Astronomy* 19 (1988), 217–237.

Dreyer, J. L. E. *A History of Astronomy from Thales to Kepler*. 2d ed. New York: Dover Publications, 1953.

———. *Tycho Brahe, A Picture of Scientific Life and Work in the Sixteenth Century*. New York: Dover Publications, 1963.

Duhem, Pierre. *To Save the Phenomena*. Tr. Edmund Doland and Chaninah Maschler. Chicago: University of Chicago Press, 1969.

Field, J. V. *Kepler's Geometrical Cosmology*. Chicago: University of Chicago Press, 1988.

Funkenstein, Amos. *Theology and the Scientific Imagination from the Middle Ages to the Seventeenth Century*. Princeton: Princeton University Press, 1986.

Gade, John A. *The Life and Time of Tycho Brahe*. New York: Greenwood Press, 1969.

Galilei, Galileo. *Dialogues Concerning Two New Sciences*. Tr. Henry Crew and Alfonso de Salvio. New York: Macmillan, 1914.

———. *Dialogues Concerning the Two Chief World Systems—Ptolemaic and Copernican*. Tr. Stillman Drake. Berkeley: University of California Press, 1953.

Gerdes, Egon W. "Kepler as Theologian." In BB, pp. 339–367.

Gilbert, W. *De Magnete*. New York: Dover Publications, 1958.

Gingerich, Owen. "Kepler as a Copernican." In *Johannes Kepler—Werk and Leistung*, pp. 109–114. Linz, 1971.

———. "Kepler's Treatment of Redundant Observations, Or, the Computer versus Kepler Revisited." In *Internationales Kepler-Symposium, Weil der Stadt 1971*, ed. Fritz Krafft, Karl Meyer, and Bernhard Sticker, pp. 307–314. Hildesheim, 1973.

———. "Kepler and the New Astronomy." *Quarterly Journal of the Royal Astronomical Society* 13 (1972), 346–373.

———. "Kepler, Johannes." In *The Dictionary of Scientific Biography*, vol. 7, ed. C. C. Gillispie and M. DeBruhl, pp. 289–312. New York: Charles Scribner's Sons, 1973.

———. "A Fresh Look at Copernicus." In *Great Ideas Today*, pp. 155–178. Chicago: Encyclopedia Britannica, 1973.

———. "From Copernicus to Kepler: Heliocentrism as Model and as Reality." In *Proceedings of the American Philosophical Society* 117 (1973), pp. 513–522.

———. "The Astronomy and Cosmology of Copernicus." In *Highlights of Astronomy*, ed. Contopoulos (1974), pp. 67–85.

———. "Crisis versus Aesthetic in the Copernican Revolution." In *Vistas in Astronomy* 17, ed. A. Beer and K. Strand, pp. 85–93. Oxford: Pergamon Press, 1975.

———. "Kepler's Place in Astronomy." In BB, pp. 261–278.

———. "The Origin of Kepler's Third Law." In BB, pp. 595–601.

———. "Ptolemy, Copernicus, and Kepler." In *The Great Ideas Today*, ed. Mortimer J. Adler and John van Doren, pp. 137–180. Chicago: Encyclopedia Britannica, 1983.

———. "Kepler's Anguish and Hawking's Query: Reflections on Natural Theology." Paper presented at the Center of Theological Inquiry, Princeton, New Jersey, November 14, 1990.

———. "Kepler, Galilei, and the Harmony of the World." In *Ptolemy, Copernicus, and Kepler*. New York: American Institute of Physics, 1991.

Godwin, Joscelyn. *Robert Fludd*. Boulder, Colorado: Shambhala Publications, 1979.

Grant, Edward. "Late Medieval Thought, Copernicus, and the Scientific Revolution." *Journal of the History of Ideas* 23 (1962), 197–220.

Guerlac, Henry. "Copernicus and Aristotle's Cosmos." *Journal of the History of Ideas* 29 (1968), 109–113.

Haase, Rudolph. "Kepler's Harmonies between *Pansophia* and *Mathesis Universalis*." In BB, pp. 519–533.

Hanson, Norwood. "The Logic of Discovery." *Journal of Philosophy* 55 (1958), 1073–1089. Reprinted in *Science, Methods, and Goals*, ed. B. Brody and N. Capaldi. New York: Benjamin, 1968.

———. "The Copernican Disturbance and the Keplerian Revolution."

Journal of the History of Ideas 22 (1961), 169–184.

———. "Is There a Logic of Scientific Discovery?" In *Current Issues in the Philosophy of Science*, ed. H. Feigl and G. Maxwell. New York: Holt, Rinehart, and Winston, 1961.

———.*Patterns of Discovery*. Cambridge: Cambridge University Press, 1961.

———. "Contra-Equivalence: A Defense of the Originality of Copernicus." *Isis* 55 (1964), 308–325.

———. *Constellations and Conjectures*, ed. Willard C. Humphreys. Dordrecht: Reidel, 1973.

Hellman, Doris. "Kepler and Tycho Brahe." In BB, pp. 223–229.

Helm, Eugene. "The Vibrating String of the Pythagoreans." *Scientific American* 217 (1967), 93–103.

Holton, Gerald. "Johannes Kepler's Universe: Its Physics and Metaphysics." *American Journal of Physics* 24 (1956), 340–351. Reprinted in *Toward Modern Science* 2, ed. Robert M. Palter. New York: The Noonday Press, 1961. Reprinted again in *Thematic Origins of Scientific Thought: Kepler to Einstein*, pp. 69–90. Cambridge, Mass.: Harvard University Press, 1974.

———. "Johannes Kepler: A Case Study in the Interaction between Science, Metaphysics, and Theology." *Philosophical Forum* 21 (1956), 21–33.

Hopkins, Jasper. *A Concise Introduction to the Philosophy of Nicholas of Cusa*. 2d ed. Minneapolis: University of Minnesota Press, 1980.

Hübner, Jürgen. "Johannes Kepler als theologischer Denker." In *Festschrift 1971*, ed. Ekkehard Preuss, pp. 21–44. Regensburg, 1971.

———. "Naturwissenschaft als Lobpreis des Schöpfers-Theologische Aspekte der naturwissenschaftlichen Arbeit Keplers." In *Internationales Kepler-Symposium, Weil der Stadt 1971*, ed. Fritz Krafft, Karl Meyer, and Bernhard Sticker, pp. 307–314. Hildesheim, 1973.

———. *Die Theologie Johannis Keplers zwischen Orthodoxie und Naturwissenschaft*. Tübingen, 1975.

———. "Kepler's Praise of the Creator." In BB, pp. 369–382.

Jammer, Max. *Concepts of Mass*. Cambridge, Mass.: Harvard University Press, 1961.

———. *Concepts of Force*. New York: Harper Torchbooks, 1962.

Jardine, Nicholas. *The Birth of History and Philosophy of Science: Kepler's "A Defense of Tycho against Ursus."* Cambridge: Cambridge University Press, 1984.

Kennedy, E. S. "Late Medieval Planetary Theory." *Isis* 57 (1966), 365–378.

Kepler, Johannes. *Joannis Kepleri Astronomi Opera Omnia*. Ed. C.

Frisch. 8 vols. Frankfurt and Erlangen, 1858–1871. Reprinted, Hildesheim, 1971–.

———. *Gesammelte Werke*. Ed. Von Dyck, Max Caspar, F. Hammer, and M. List. Munich, 1937–.

———. *Johannes Kepler in seinen Briefen*. Vols. 1 and 2. Ed. Max Caspar and W. von Dyck. Munich and Berlin: R. Oldenbourg, 1930.

———. *Kepler's Conversation with Galileo's Sidereal Messenger*. Tr. Edward Rosen. New York: Johnson Reprint, 1965.

———. *Kepler's Dream*. Ed. John Lear. Tr. Patricia Kirkwood. Berkeley: University of California Press, 1965.

———. *The Six-Cornered Snowflake*. Ed. and tr. Colin Hardie. Oxford: Clarendon Press, 1966.

———. *Kepler's Somnium*. Tr. Edward Rosen. Madison: University of Wisconsin Press, 1967.

———. *On the More Certain Fundamentals of Astrology*. Tr. Mary Ann Rossi. Appleton, Wisconsin, 1976.

———. *Mysterium Cosmographicum*. Tr. A. M. Duncan. New York: Abaris Books, 1981.

———. *Astronomia Nova*. Tr. Owen Gingerich and William Donahue. In *The Great Ideas Today*, ed. Mortimer J. Adler and John van Doren, pp. 305–341. Chicago: Encyclopedia Britannica, 1983.

———. *Epitome of Copernican Astronomy*, books 4 and 5. Tr. Charles Glenn Wallis, *Great Books of the Western World* 16, pp. 839–1004. Chicago: Encyclopedia Britannica, 1984.

———. *The Harmonies of the World*, book 5. Tr. Glenn Wallis, *Great Books of the Western World* 16, pp. 1005–1085. Chicago: Encyclopedia Britannica, 1984.

———. *The New Astronomy*. Tr. William Donahue. Cambridge: Cambridge University Press, 1992.

Kleiner, Scott A. "A Look at Kepler and Abductive Argument." *Studies in History and Philosophy of Science* 14 (1983), 289–313.

Koestler, Arthur. *The Sleepwalkers*. New York: The Macmillan Co., 1959.

———. *The Watershed*. New York: Anchor Books, 1960.

———. "Kepler and the Psychology of Discovery." In *The Logic of Personal Knowledge*, pp. 49–57. London, 1961.

Koyré, Alexandre. "Influence of Philosophic Trends on the Formulation of Scientific Theories." In *The Validation of Scientific Theories*, ed. P. Frank, pp. 177–187. New York: Collier Books, 1961.

———. *Newtonian Studies*. Chicago: University of Chicago Press, 1968.

———. *From the Closed World to the Infinite Universe*. Baltimore: The Johns Hopkins Press, 1968.

———. *The Astronomical Revolution.* Tr. R. E. W. Maddison. Ithaca, New York: Cornell University Press, 1973.

Krafft, Fritz. "Copernicus and Kepler: New Astronomy from Old Astronomy." In BB, pp. 287–306.

———. "The New Celestial Physics of Johannes Kepler." In *Physics, Cosmology, and Astronomy, 1300–1700: Tension and Accommodation,* ed. Sabetai Unguru, pp. 185–227. Dordrecht: Kluwer Academic Publishers, 1991.

Kuhn, Thomas. *The Copernican Revolution.* New York: Vintage Books, 1959.

Langford, Jerome J. *Galileo, Science, and the Church.* Rev. ed. Ann Arbor: University of Michigan Press, 1971.

Lindberg, David. "Kepler and the Incorporeality of Light." In *Physics, Cosmology, and Astronomy, 1300–1700: Tension and Accommodation,* ed. Sabetai Unguru, pp. 229–250. Dordrecht: Kluwer Academic Publishers, 1991.

Lindberg, David, ed. *God and Nature: Historical Essays on the Encounter between Christianity and Science.* Berkeley: University of California Press, 1986.

Lindberg, David, and Robert Westman, eds. *Reappraisals of the Scientific Revolution.* Cambridge: Cambridge University Press, 1990.

List, Martha. "Kepler as a Man." In BB, pp. 97–105.

———. "Kepler und die Gegenreformation." In *Kepler Festschrift, 1971,* pp. 45–63. Regensburg, 1971.

Losse, J. *A Historical Introduction to the Philosophy of Science.* Oxford: Oxford University Press, 1972.

McMullin, Ernan. "The Conception of Science in Galileo's Work." In *New Perspectives on Galileo,* ed. R. E. Butts and J. C. Pitts, pp. 209–257. Dordrecht: Reidel, 1978.

Mittelstrass, Jürgen. "Methodological Elements of Keplerian Astronomy." Tr. J. Fearns, *Studies in History and Philosophy of Science* 3 (1972), 203–232.

Moesgaard, Kristian. "Copernican Influence on Tycho Brahe." In *The Reception of Copernicus' Heliocentric Theory,* ed. Jerzy Dobrzycki, pp. 31–55. Dordrecht: Reidel, 1972.

Monk, Robert. "The Logic of Discovery." *Philosophy Research Archives* 3 (1977), A_10–E_6.

Moody, Ernest, and Marshall Clagett. *The Medieval Science of Weights.* Madison: University of Wisconsin Press, 1960.

Neugebauer, O. "On the Planetary Theory of Copernicus." In *Vistas in Astronomy* 10, ed. Arthur Beer, pp. 89–103. Oxford: Pergamon Press, 1968.

Pauli, Wolfgang. "The Influence of Archetypal Ideas in the Scientific Theories of Kepler." Tr. Priscilla Silz, in *Interpretation of Nature and of the Psyche*, pp. 151–247. New York: Pantheon Books, 1955.

Peirce, Charles. *Selected Writings*. Ed. Philip Wiener. New York: Dover, 1966.

Petri, Winfried. "Die betrachtende Kreatur im trinitarischen Kosmos. Auswahl aus Buch 1 und Buch 4, Teil 1, von Johannes Keplers 'Epitome Astronomiae Copernicanae.'" In *Kepler Festschrift 1971*, ed. Ekkehard Preuss, pp. 64–98. Regensburg, 1971.

Popper, Karl. *The Logic of Scientific Discovery*. 2d ed. New York: Harper Torchbooks, 1968.

Preuss, Ekkehard, ed. *Kepler Festschrift 1971*. Regensburg, 1971.

Price Derek. "Contra–Copernicus: A Critical Reestimation of the Mathematical Planetary Theory of Ptolemy, Copernicus, and Kepler." In *Critical Problems in the History of Science*, ed. Marshall Clagett, pp. 197–218. Madison: University of Wisconsin Press, 1969.

Ptolemy, C. *The Almagest*. Tr. Catesby Taliaferro, in *Ptolemy, Copernicus, Kepler, Great Books of the Western World* 16, pp. 1–478. Chicago: Encyclopedia Britannica, 1982.

Rosen, Edward. "Kepler's Defense of Tycho against Ursus." *Popular Astronomy* 54 (1946), 405–412.

———. "Maestlin, Michael." In *Dictionary of Scientific Biography*, vol. 9, ed. C. G. Gillispie, pp. 167–170. New York: Charles Scribner's Sons, 1973.

———. "Kepler and the Lutheran Attitude." In BB, pp. 317–337.

———. "Kepler's Place in the History of Science." In BB, pp. 279–285.

———. *Three Imperial Mathematicians; Kepler Trapped between Tycho Brahe and Ursus*. New York: Abaris, 1986.

Rosen, Edward, ed. *Three Copernican Theses*. New York: Octagon Books, 1971.

Russell, J. L. "Kepler's Laws of Planetary Motion: 1609–1666." *British Journal for the History of Science* 2 (1964), 1–24.

Salisbury, Thomas. *Mathematical Collections*. London, 1661.

Schedl, Claus, "Die logotechnische Struktur des Psalms VIII und Keplers Weltharmonik." In *Johannes Kepler: 1571–1630, Gedenkschrift der Universität Graz*, pp. 105–123. Graz, 1975.

Seeger, Raymond. "On Kepler as a Physicist." In BB, pp. 693–698.

Shapere, Dudley. "Descartes and Plato." *Journal of the History of Ideas* 24 (1963), 572–576.

———. *Galileo*. Chicago: University of Chicago Press, 1974.

———. "Copernicanism as a Scientific Revolution." In *Copernicanism Yesterday and Today*, ed. K. Beer and K. Strand, pp. 95–104. New

York: Pergamon, 1975.

Small, Robert. *An Account of the Astronomical Discoveries of Kepler.* Madison: University of Wisconsin Press, 1963.

Stephenson, Charles Bruce. *Kepler's Physical Astronomy.* New York: Springer–Verlag, 1987.

Straker, Stephen M. "Kepler's Optics: A Study in the Development of 17th-Century Natural Philosophy." Ph.D. Dissertation, Indiana University, 1971.

Strong, Edward. *Procedures and Metaphysics.* Hildesheim: Georg Olms, 1966.

Suppe, Frederick. *The Structure of Scientific Theories,* 2d ed. Urbana: University of Illinois Press, 1977.

Trinkhaus, Charles. *In Our Image and Likeness: Humanity and Divinity in Italian Humanist Thought.* 2 vols. London: Constable, 1970.

Werner, Eric. "The Last Pythagorean Musician: Johannes Kepler." In *Aspects of Medieval and Renaissance Music,* ed. Jan LaRue, pp. 867–882. New York, 1966.

Westfall, Richard. *Science and Religion in Seventeenth-Century England.* New Haven: Yale University Press, 1958.

Westman, Robert. "Johannes Kepler's Adoption of the Copernican Hypothesis." Ph.D. Dissertation, University of Michigan, 1971.

———. "The Comet and the Cosmos: Kepler, Maestlin, and the Copernican Hypothesis." In *The Reception of Copernicus' Heliocentric Theory,* ed. Jerzy Dobrzycki, pp. 7–30. Dordrecht: Reidel, 1972.

———. "Kepler's Theory of Hypothesis and the 'Realist Dilemma.'" *Studies in History and Philosophy of Science* 3 (1972), 233–264.

———. ed. *The Copernican Achievement.* Berkeley: University of California Press, 1975.

Whewell, William. *History of the Inductive Sciences,* vol. 1. New York: Appleton, 1859.

White, Andrew. *A History of the Warfare of Science with Theology in Christendom.* New York: George Braziller, 1955.

Whiteside, D. T. "Keplerian Planetary Eggs, Laid and Unlaid, 1600–1605." *Journal for the History of Astronomy* 5 (1974), 1–21.

Wilson, Curtis. "Kepler's Derivation of the Elliptical Path." *Isis* 59 (1968), 5–25.

———. "From Kepler's Laws, so-called, to Universal Gravitation: Empirical Factors." *Archive for History of Exact Sciences* 6 (1970), 89–170.

———. "How Did Kepler Discover His First Two Laws?" *Scientific American* 226 (1972), 92–106.

———. "The Inner Planets and the Keplerian Revolution." *Centaurus*

17 (1973), 205–248.

———. "Newton and Some Philosophers on Kepler's Laws." *Journal of the History of Ideas* 35 (1974), 231–258.

———. "Kepler's Ellipse and Area Rule—Their Derivation from Fact and Conjecture." In BB, pp. 587–591.

———. "Rheticus, Ravetz, and the 'Necessity of Copernicus' Innovation." In *The Copernican Achievement*, ed. Robert Westman, pp. 17–39. Berkeley: University of California Press, 1975.

———. "Horrocks, Harmonies, and the Exactitude of Kepler's Third Law." In *Science and History*, vol. 16 of *Studia Copernicana*, pp. 235–259. Ossolineum, 1978.

INDEX

ABOUT THE AUTHOR

Job Kozhamthadam, S.J., received his Ph.D. from the University of Maryland, College Park. He is currently Reader in Philosophy of Science at the Pontifical Athenaeum, Pune, India, as well as Visiting Associate Professor of Philosophy of Science at Loyola University of Chicago. In addition to *The Discovery of Kepler's Laws: The Interaction of Science, Philosophy, and Religion*, Kozhamthadam has published many articles on the interaction between science and religion.